Net-Zero and Low Carbon Solutions for the Energy Sector

Net-Zero and Low Carbon Solutions for the Energy Sector

A Guide to Decarbonization Technologies

Amin Mirkouei
University of Idaho, Idaho Falls, ID, USA

Published by John Wiley & Sons, Inc., Hoboken, New Jersey.
Published simultaneously in Canada.

For general information on our other products and services or for technical support, please contact our Customer Care Department within the United States at (800) 762-2974, outside the United States at (317) 572-3993 or fax (317) 572-4002.

Wiley also publishes its books in a variety of electronic formats. Some content that appears in print may not be available in electronic formats. For more information about Wiley products, visit our web site at www.wiley.com.

Library of Congress Cataloging-in-Publication Data Applied for:

Hardback ISBN: 9781119982166

Cover Design: Wiley
Cover Image: © Andriy Onufriyenko/Getty Images

Set in 9.5/12.5pt STIXTwoText by Straive, Chennai, India

SKY10067419_021724

Contents

About the Author

Dr. Amin Mirkouei is an associate professor at the University of Idaho, certified professional engineer (PE), and an experienced technologist. He has over 10 years of experience contributing and leading cross-disciplinary projects in decarbonization technologies, renewable materials, sustainable energy systems, design and manufacturing, cyber-physical control and optimization, and operations research, particularly renewable fuels, green chemicals, and rare earth elements and minerals from various waste streams, such as biomass feedstocks, plastic wastes, e-wastes, and animal manure. Currently, he is a major advisor in Industrial Technology, Mechanical Engineering, Biological Engineering, and Environmental Science programs at the University of Idaho in Idaho Falls, where he directs the Renewable and Sustainable Manufacturing Laboratory (RSML). RSML aims to maintain many research opportunities that can positively impact all segments of sustainable manufacturing, especially sustainable food–energy–water systems (FEWS).

Dr. Mirkouei has served on several university committees, such as member of the UI President's Sustainability Working Group, UI Safety and Loss Control Committee, and chair of the UI-IF Environmental, Health, and Safety Committee. He also served in ASME conference as organizer, Forbes as sustainability contributor, several journals as editorial board member or reviewer, and several federal agency panels as a proposal reviewer (e.g., NSF SBIR, USDA-NIFA, and NSF ERC programs). He received several state and national media coverages in Here We Have Idaho, Yahoo Finance, Bloomberg, etc. about several of his projects.

In addition, he has published and co-authored over 40 articles in scientific journals and peer-reviewed conference proceedings. He also received over $2.5 million in research grants from private companies and state and federal agencies, as well as honors and awards, such as the "2022 University of Idaho Interdisciplinary and Collaboration Excellence Award" and the 2022 ASME/IDETC-DFMLC best paper award.

Here is the list of his sponsors and collaborators:

- Idaho Global Entrepreneurial Mission (IGEM)-Commerce
- Aquaculture Research Institute (ARI)
- Idaho Water Resources Research Institute (IWRRI)
- Idaho Geological Survey (IGS)
- United States Geological Survey (USGS)
- University of Idaho Office of Research and Economic Development (ORED)
- Center for Advanced Energy Studies (CAES)
- National High Magnetic Field Laboratory (MagLab)
- Idaho National Laboratory (INL)
- College of Eastern Idaho (CEI)
- National Science Foundation (NSF)
- Idaho Strategic Resources Inc. (IDR)
- Riverence Provisions LLC

You can find more information about the author and his projects at the RSML website: https://webpages.uidaho.edu/rsml.

Acknowledgments

I am inspired by earlier research and studies in renewable and sustainable energy, and I have been fortunate to work with some leading researchers and experts in this field. I sincerely thank all my students and others who contributed to this book or offered advice and suggestions. I am also very grateful for the support from Wiley, especially Summers Scholl, Executive Editor of Physical Sciences, Kubra Ameen, Managing Editor, and Hafiza Tasneem, Content Refinement Specialist, at Wiley. Finally, I would like to appreciate readers' new ideas, comments, and recommendations, and I will strive to address them in future editions.

Acronyms

ATR	Autothermal reforming
a-Si:H	Hydrogenated amorphous silicon
BOS	Balance of system
BTU	British Thermal Unit
CCUS	Carbon capture, utilization, and sequestration
C-Si	Crystalline silicon
CdTe	Cadmium telluride
CFC	Chlorofluorocarbon
CHP	Combined heat and power
CIGS	Copper indium gallium diselenide
CPV	Concentrated photovoltaic
CPVT	Concentrated photovoltaic with thermal systems
CSP	Concentrated solar power
DOE	U.S. Department of Energy
EOR	Enhanced oil recovery
EGS	Enhanced geothermal systems
EIA	U.S. Energy Information Administration
FAME	Fatty acid methyl esters
FT	Fischer–Tropsch
GE	General Electric
GHG	Greenhouse gas
GWdc	Gigawatts direct current
GWP	Global warming potential
HCFC	Hydrochlorofluorocarbon
IAEA	International Atomic Energy Agency
IEA	International Energy Agency
IGCC	Integrated gasification combined cycle
IPCC	Intergovernmental Panel on Climate Change
kWh	Kilowatt hour

LCOE	Levelized cost of energy
LED	Light-emitting diode
LFR	Linear Fresnel reflector
MeO	Metal oxide
MGS	Metallurgical-grade silicon
MWth	MW thermal energy
NGCC	Natural gas combined cycle
NETL	National Energy Technology Laboratory
NREL	National Renewable Energy Laboratory
ORC	Organic Rankine cycle
PCM	Phase-change material
PEM	Polymer electrolyte membrane
PET	Polyethylene terephthalate
PPM	Parts per million
PV	Photovoltaic
PVF	Polyvinyl fluoride
PVDF	Polyvinylidene fluoride
PVT	Photovoltaic with thermal systems
SCPC	Supercritical pulverized coal with carbon capture
SFR	Sodium-cooled fast reactor
SNG	Synthetic natural gas
SMR	Stream methane reforming
TRL	Technology readiness level
TWh	Terawatt hours
USGS	United States Geological Survey
WEF	World Economic Forum
WGS	Water–gas shift

List of Figures

List of Tables

Introduction

The climate impacts of rising global temperatures to 2 °C and beyond are disastrous. The evidence is all around us, such as rising sea levels, warming oceans, melting glaciers, and more frequent and intense floods, fires, droughts, storms, and heat waves. According to the Intergovernmental Panel on Climate Change (IPCC), the current emission reduction efforts to keep warming at 1.5 °C are nowhere near enough. Recent IPCC reports show that energy consumption, deforestation and land use, and industrial chemicals and cement are the major greenhouse gas (GHG) emission contributors, with around 70, 10, and 3%, respectively. Approximately 59% of the emission remains in the air, and 41% is captured by nature's reservoirs, such as land and ocean, with around 24 and 17%, respectively. Latest studies estimated that human activities annually release over 50 billion metric tons of GHGs (including CO_2, CH_4, and N_2O), of which CH_4 has over 30% global warming potential (GWP) in the short term (20–50 years) and CO_2 has a greater GWP in the long term (100 years).

We started adding emissions to the atmosphere unintentionally about 300 years ago, which resulted in increasing the atmospheric CO_2 concentration (around 120 ppm) and changing our climate. According to the IPCC reports, there is a 66% chance of keeping the global temperature below 2 °C if no more than 1,000 billion tons of GHG are emitted between 2011 and 2100, and since 2011, 200 billion tons have been emitted. The 2015 Paris Agreement concluded that fully implemented strategies and policies after 2030 will lead to 2.7 °C by 2100. In response, decarbonization technologies can reduce GHG emissions and limit the rising global temperature below 2 °C by taking GHGs from the atmosphere and putting them back into geologic reservoirs and terrestrial ecosystems.

Most climate studies and assessment models reported that CO_2 concentration must start reducing (or stop increasing) in the 21st century to meet the 2 °C target and the associated climate crises, and address the raised 1 °C in the 20th century. We have two types of emissions: biogenic (natural-made) and anthropogenic (human-caused). The dominant anthropogenic sources are energy consumption, agriculture, land-use change, and cement production. Businesses generate direct and indirect anthropogenic emissions. Direct emission sources include business-owned facilities and equipment, such as vehicles. Indirect emission sources include purchased resources (water and electricity for heating and cooling) and equipment and facilities that are not owned by the business but that the company uses for various purposes and services, such as supply chain and distribution. According to the Energy and Climate Intelligence Unit and Oxford Net Zero, roughly 20% (400 out of 2,000) of the world's largest publicly traded companies have established net-zero targets, and roughly 30% (120 out of 400) of these companies aim to achieve these targets by 2030.

Companies must shift their operations, use carbon removal technologies, support climate solutions and policies, prioritize suppliers that adopt emission reduction targets, support low-carbon materials, phase out unsustainable operations, engage employees on climate solution opportunities, and offer them climate-friendly investments. The latest report from IPCC demonstrated that human activities profoundly impact environmental degradation, and global action is needed to prevent further loss and threats to biodiversity, natural carbon sinks, and ecosystems. Natural carbon sinks, such as oceans, wetlands, and forests, naturally manage the global carbon balance. Reducing supply chain emissions requires collaboration between companies and countries since it happens outside of companies and is out of their control due to the overlap. According to the World Economic Forum (WEF) report, roughly 83% of carbon footprint release is outside of a company's direct control and considered supply chain emissions. To fully control the products' life cycle, businesses must avoid or limit partnering with or serving fossil fuel companies that hurt global efforts to reduce carbon emissions as much as possible. Wealthier companies carry more responsibilities than emerging companies due to their higher emissions and footprints from current and past business developments. Businesses can prioritize sustainable waste management, carbon-friendly production, low-carbon materials, and circular practices by producing their products from reused and recycled materials.

Nature-based solutions can reduce climate-related events (e.g., droughts, floods, fires, and resource scarcity) by around 37% through protecting and restoring forests and wetlands, and properly managing agriculture and grasslands. Electrification powered by renewable energy sources in some energy-intensive sectors will be the heart of the energy transition and crucial to net-zero emissions success. The core sources are biomass, solar, onshore and offshore wind, and geothermal. Various sustainability and circular economy methods, such as carbon capturing, utilization, and sequestration (CCUS) and 6R (i.e., reduce, reuse, recycle, recover, redesign, and remanufacturing), offer a comprehensive response to climate-related crises by mitigating pollution and waste. The 2021 global economy was around 8.5% circular, which required more attention on the segments and steps that consume more resources. For example, 6R methods can tackle emissions (over 45%) associated with making products by reducing the demand for raw materials, such as plastic, cement, aluminum, or steel. The emphasis on 6R and circular economy methods is simply for overshooting emissions earlier and avoiding environmental consequences that cannot be fully addressed by future GHG removal, and can lead to crucial irreversible crises, such as land degradation and biodiversity loss.

The long-term goal of net-zero (carbon neutral) emission target is to keep the increased warming at 1.5 °C by 2050, using different net-zero and low-carbon solutions in various sectors. However, the net-zero target only works if all companies and countries commit the same, which is a highly unlikely prospect. In addition, net-zero targets have several unintended consequences (biodiversity loss) due to delaying emission reduction efforts. Therefore, we must pursue decarbonization solutions (e.g., CCUS technologies and forest restoration and protection) and net-negative targets to address the lack of accountability mechanisms. Decarbonization technologies strive to remove more emissions than they emit to meet intermediate target (net-zero) goals.

Several businesses are not able to meet net-zero targets, such as fossil fuel producers, airlines, and steel manufacturers; however, they pay other companies to remove emissions from the air (carbon offsetting), using different solutions, such as renewable material development or reforestation. Carbon offsetting is one of the emission reduction strategies for companies that cannot change their entire operations and reduce their emissions to zero in the short term due to the lack of technologies or pathways to remove emissions at a meaningful scale. While carbon offsetting has benefits to protecting the natural ecosystems, it can host a lot of problems in other locations by granting a few companies permission to avoid shifting their business models. Several operations can emit GHG emissions or heat-trapping gases (e.g., CO_2, CH_4, and N_2O), and pollutants into the atmosphere, such as traveling by vehicles and airplanes, burning fossil fuels for power generation, transportation, cooling, and heating, and manufacturing metals and cement, degrading soils, forests, and other ecosystems. Most of these heat-trapping gases stay in the Earth's atmosphere, a portion of these emissions is captured through natural processes, such as photosynthesis by plants in soils, seas, or oceans.

Decarbonization and carbon capturing strategies are the points that we can reduce GHG emissions in the atmosphere and eventually limit the warming. The Earth is warming up due to natural and anthropogenic GHGs that can trap heat. These emissions have different attributes and lifetimes: some of them can trap way more heat (e.g., CH_4, N_2O, CFC, and HFC), and some of them last longer in the atmosphere, such as CO_2. Particularly, CO_2 stays for centuries in the atmosphere, but CH_4 remains for decades. The concept of carbon budget introduced by IPCC is the total emissions (including from GHG and land-use change) that can be added without exceeding the target temperatures. Based on the IPCC, the budget limit for the 2 °C target is 1,000 billion tons of CO_2 by 2100.

Over 90% of GHGs come from primary, secondary, and tertiary sectors of economy, particularly energy, agriculture and land use, manufacturing, transportation, and construction. Power generation is the biggest contributor, emitting roughly 25% of GHGs from burning natural gas or coal at power plants. The 2nd biggest contributor is agriculture and land use for food production, emitting around 24% of GHGs. Manufacturing industry creates approximately 21% of GHGs for producing different products, such as plastics, cement, and steel. Transportation and building generate roughly 14 and 6% of GHGs, respectively. The last 10% GHG emitters are mainly from escaped GHGs accidentally from the energy sector, such as leaky oil and natural gas pipelines. Finally, attacking the five main GHG emitters is the 1st step to meeting low-carbon economy and eventually net-zero targets. These emission contributors are interconnected to one another; for example, we use power and industrial products (e.g., furnaces, air conditioners, and heaters) in our buildings.

Recent studies reported that 12% of generated power is consumed in buildings for lighting, heating, and cooling, and 11% is used in industry for manufacturing different products. For every kilowatt hour (kWh) of power generation from coal and natural gas, the emitted CO_2 is roughly 1 and 0.5 kg (2.2 and 1.2 lbs), respectively. Other sources (e.g., wind, solar, and nuclear) do not emit any CO_2 during power generation. An average American house consumes around 1,100 kWh per month, which is roughly 7,700 kg (17,000 lbs) of CO_2 per year. Consuming power more efficiently and shifting to renewable sources can reduce the emitted CO_2 to around zero. There are several solutions to enhance efficiency, using new technologies and automation, such as light-emitting diode (LED) lights, smart thermostats and heating systems, efficient insulations and roofs, and high-performance glasses and pumps. To shift power generation from coal and natural gas, we can use solar panels, wind turbines, hydroelectric, geothermal, biomass-based energy, nuclear power, and waste-based energy from methane capturing.

Manufacturing industry produces many products, using big machines and intensive operations with high temperatures. The main products that emit the most GHGs are metals (5%), chemicals (3%), cement (3%), and waste (3%). Roughly, 1–1.5% of GHGs come from plastic production, use, and disposal. To meet net-zero targets, we must improve processes, materials, and waste management in lands and water resources. Also, we should transition to climate-friendly refrigerants to limit the use of chlorofluorocarbons (CFCs), hydrochlorofluorocarbons (HCFCs), and other chemicals that can contribute to global warming and damage the ozone layers. Other solutions are using waste for renewable energy and recycled material production.

The major GHG emitter in the transportation sector is road transport (with roughly 10%), using personal vehicles or trucks. Commercial and military aviation is the 2nd major contributor to GHG, with roughly 2%. For example, burning a gallon of gasoline releases around 9 kg (20 Ibs) of CO_2 that could stay in the atmosphere for centuries. Roughly 5,440 kg (12,000 Ibs) of CO_2 per year is emitted by an average car with 25 miles per gallon and 15,000 miles per year. On average, airplanes get up to 60 miles per gallon per passenger. To meet the net-zero targets, we can shift to alternative options, such as walking, biking, mass transportation, and hybrid or electric vehicles, as well as enhance the efficiency of the existing cars, trucks, trains, airplanes, and ships.

Residential and commercial buildings are the primary GHG contributors, emitting roughly 4 and 2%, respectively, using furnaces, boilers, heaters, air conditioners, refrigerators, and freezers. An average home in the southern U.S. with a warm climate burns approximately 100,000 BTUs (British Thermal Unit or a therm) of natural gas for cooking or water and space heating that releases 6 kg (over 13 Ibs) of CO_2. In a colder climate, it can reach up to 1,000 therm per year, emitting 5,900 kg (over 13,000 Ibs) of CO_2.

What to Expect in This Book

This book can serve as a comprehensive desk reference for seasoned professionals, focusing on net- or near-zero solutions for the energy sector since energy generation, storage, and consumption release over 70% of GHG emissions. For those who are new in this field or support environmentally friendly solutions, this book can provide a step-by-step guide, context, and encouragement to understand the existing solutions in the energy sector. To meet net-zero targets, we need to shift to alternative energy sources that do not emit GHG emissions during generation, storage, and consumption (e.g., solar, wind, and hydropower) or use low-carbon energy sources (e.g., nuclear power and natural gas), along with enhancing efficiencies in various sectors. According to the IPCC and Global Carbon Project, current major emission contributors are power generation (25%), food, agriculture, and land use (24%), industry (21%), transportation (14%), buildings (6%), and others (10%).

This book proposes mature net-zero and low-carbon pathways and technologies in the energy sector for producing and storing power, heat, biofuel, and hydrogen. It also highlights various processes for achieving net-zero targets and addressing climate concerns, with case studies demonstrating their applications. Each chapter provides a case study, covering decarbonization solutions that have high potential to be used in the near future, such as solar hybrid systems for net-zero power generation, CCUS hybrid systems for low-carbon power generation, pumped hydropower for power storage, commercial concentrating solar power plants for heat generation, gasification with CCUS for biofuel production, and hybrid thermochemical processes for hydrogen production. It is not enough to have solutions; we need realistic, sustainable, and feasible solutions regarding scale and resources. In this book, the main focus is on solutions with high technology readiness levels (TRLs). TRL is a measure for evaluating a technology's maturity from basic research to commercialization, and a higher number indicates that the technology is closer to commercialization in the market. The standard TRL classification is as follows:

- TRL 1: Initial idea with basic concept and principles
- TRL 2: Application formulated using the solution concept
- TRL 3: Solution needed to be prototyped, applied, and validated
- TRL 4: Early prototype in test conditions
- TRL 5: Large prototype with proven components in specific conditions
- TRL 6: Full prototype at scale
- TRL 7: Precommercial demonstration in expected conditions
- TRL 8: First commercial demonstration, full-scale deployment in final form
- TRL 9: Commercial operation in a relevant environment, needs evolutionary improvement to stay competitive
- TRL 10: Integration needed at scale with further integration efforts
- TRL 11: Proof of stability reached with predictable growth

Audience

This book is written to be a valuable resource for businesses, academics, and policymakers looking for net-zero and low-carbon solutions in the energy sector and actively contributing to net-zero emission targets for keeping the atmospheric CO_2 equivalent levels below the dangerous range, for example, Environmental Scientists, Mechanical, Industrial, and Manufacturing Engineers, Chemical and Biological Engineers, Process Designers, Policymakers, Chemists and Biologists, Government Employees in Economic and Environmental Affairs, Professors, Post-Doc Scholars, Graduate and Undergraduate students in Engineering, Science, and Business Schools, Managers, CEOs, and any company in a position to invest in these technologies across industrial sectors.

Benefits of Applying this Book

Recent studies reported that the decade to 2030 will be decisive in addressing climate concerns. The existing solutions are no longer sufficient to meet net-zero emission targets and achieve the Paris Agreement goals. Global leaders and governments must commit to net-zero and net-negative (removal exceeds emissions) targets to keep the atmospheric CO_2 equivalent levels below the dangerous range of 500 parts per million (ppm) by mid-century. Currently, the global energy sector is the primary source of GHG emissions. Fossil fuels (coal, petroleum, and natural gas) production and combustion represent 89% of global GHG emissions. For rapid transition, we need to use renewable or low-carbon energy sources. Electrification powered by renewable resources (wind and solar) plays a crucial role in phasing out fossil fuels, mainly coal and petroleum. Natural gas has lower environmental impacts than coal and crude oil, which can be used for power generation before a complete transition to renewable alternatives. Steel, cement, and chemical manufacturing, as well as long-distance and heavy transportation, are some of the most energy-intensive economic activities, which are not yet technically feasible for electrification powered by renewable or low-carbon energy sources. Meeting ambitious international climate goals may require global carbon emissions to fall below zero in the second half of this century, achieving what is known as net-zero emissions. Carbon capture, utilization, and sequestration strategies are not only a long-term solution. The technologies can also play an essential, near-term role in clean-energy transitions. They can neutralize or offset emissions that are currently technically challenging or prohibitively expensive to address.

1

Power Generation (Net-Zero Solutions)

Energy is the most important source for economic growth and food-water security of a nation. The primary energy sources on the earth are the sun, geothermal, nuclear reactions, fossil fuels, and gravitational (motion) of the sun, moon, and earth (e.g., wind and tidal). Currently, most of the generated energy sources utilize fossil fuel resources (over 80% in 2022), involving hazardous gases and toxic emissions, such as CO_2, CH_4, and N_2O. Combusting fossil fuel sources has changed the chemical composition of the atmosphere and oceans, particularly increasing CO_2 in the atmosphere (from 270 to 410 ppm in 300 years) and changing the ocean pH and carbonic acid, which led to several climate-related effects, such as global warming and ocean acidification. Among greenhouse gas (GHG) emissions, CO_2 has the largest impact on the earth's climate by trapping thermal radiation into the atmosphere, raising the temperature, melting ice caps, and rising oceans. Currently, the global energy sector is the primary contributor to GHG emissions. Particularly, fossil fuel-based energy production and combustion, from coal, petroleum, and natural gas, represent 89% of global GHG emissions [1, 2]. For rapid transition, we need to use renewable or low-carbon energy sources. Electrification powered by renewable resources (e.g., wind and solar) plays a key role in phasing out fossil fuels (mainly coal and petroleum). Natural gas has lower environmental impacts than coal and oil, which can be used for power generation before a complete transition to renewable alternatives (Figure 1.1) [3]. Steel, cement, and chemicals manufacturing, as well as long-distance and heavy transportation, are some of the most energy-intensive economic activities, which are not yet technically feasible to use electrification powered by renewable or low-carbon energy sources.

Renewable energy sources are able to address energy needs and environmental degradation due to the use of fossil fuel-based energy. Among these sources, the leading renewable energy sources are the sun, geothermal, and planetary motion. Recently, the utilization of renewable or low-carbon energy sources, especially power generation from solar, wind, hydro, biomass, and nuclear, has

Net-Zero and Low Carbon Solutions for the Energy Sector: A Guide to Decarbonization Technologies, First Edition. Amin Mirkouei.
© 2024 John Wiley & Sons, Inc. Published 2024 by John Wiley & Sons, Inc.

Figure 1.1 Energy generation from renewable resources. Source: John Wiley & Sons.

attracted huge interest. These energy sources can meet various energy needs in different sectors, such as agriculture, manufacturing, and construction (housing), along with addressing sustainability challenges, such as socio-environmental aspects, including pollution and job creation. In order to be economically viable, supportive policies, such as carbon tax, subsidies, or feed-in tariff, are necessary. According to the U.S. Energy Information Administration (EIA), the total energy consumption will increase to nearly 50% by 2050 [4], and the United States recently committed to reducing 50% of the GHG emissions from the 2005 level by 2030, power emission-free by 2035, and net-zero emission economy-wide by 2050 (Table 1.1) [6].

Power (electricity) generation is the most critical energy need in the world that can be evaluated using various criteria, such as science, functionality, usability,

Table 1.1 CO_2 emissions from the energy sector (million metric tons).

Source	2021	2022	2023
Coal	1,002	943	886
Natural gas	1,657	1,739	1,678
Petroleum	2,234	2,282	2,267

Source: Adapted from U.S. EIA [5].

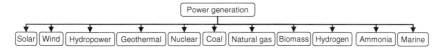

Figure 1.2 Key power generation methods.

compatibility, security, performance, sustainability (e.g., techno-economic and socio-environmental), and technology readiness level. The basic science of power generation technologies is an essential factor to consider, which includes the principles of thermodynamics, fluid mechanics, and materials science. A couple of critical parameters for power generation include reliability and efficiency, compatibility with the required configurations, security, ease of use and user-friendly interfaces, and repair and maintenance aspects. The main performance factors are efficiency, capacity, and power output which can reduce total cost and environmental impacts. Technology readiness level is a measure for evaluating a technology's maturity from basic research to commercialization, and a higher number indicates that the technology is closer to commercialization in the market. Figure 1.2 presents the commercialized methods for power generation.

An overview of key net-zero solutions for power generation using various resources is presented in this chapter. Particularly, this chapter provides 29 technologies with high technology readiness levels (TRLs), along with a case study about solar hybrid systems. Table 1.2 presents the unsubsidized levelized cost of energy (LCOE) generation, using both conventional and renewable sources [7].

1.1 Solar

The sun is the most abundant energy source for the earth and shows immense potential to satisfy global energy needs. Particularly, solar energy falls on the earth's surface at the rate of 120 petawatts (1 million gigawatts) that can meet the global demand for over 20 years with just a day's energy received from the sun [8]. Solar-based energy is one of the dominant renewable energy sources that can generate clean energy from light to voltage by harvesting photons from the sun, energizing electrons, and creating the electrical current, using photovoltaic (PV) systems (e.g., rooftop solar panels), solar thermal systems, or mixed solar PV and thermal systems (Figure 1.3). According to the International Energy Agency (IEA), solar-based energy can be a dominant source of power and heat generation and supply between 17 and 27% of the global power for keeping the average global temperature below 2°C. Globally, the primary methods use PV or solar thermal systems with either concentrated or non-concentrated collectors within

Table 1.2 Unsubsidized levelized cost of energy.

Solutions		Cost ($/MWh)
Coal		65–159
Gas combined cycle		44–73
Gas peaking		151–198
Geothermal		59–101
Nuclear		129–198
Solar	PV community	63–94
	PV crystalline	31–42
	PV rooftop	74–227
	PV thin film	29–38
	Thermal tower with storage	126–156
Wind	Onshore	26–54
	Offshore	86

Source: Adapted from Lazard [7].

solar-only and solar hybrid configurations. Latest studies reported that generating power and heat, using solar-based technologies is a sustainable and cost-effective approach to invest in various scales (e.g., small, medium, and large).

Over the last 20 years, solar power generation has grown from prototype to mature technology and reached high TRLs, but it needs further integration efforts. The latest U.S. Department of Energy (DOE) goal is to supply at least 40% of power through solar-based plants by 2035 without increasing the price [9]. The

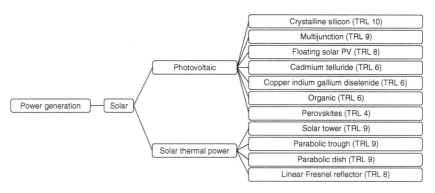

Figure 1.3 Major solar technologies for power generation.

latest studies reported that the solar power and heat industry created over 230,000 jobs with a wage higher than the national average in the United States, and the power sector could employ between 0.5 and 1.5 million people by 2035 [6]. The main demands for solar-based energy are power, space heating, hot water, space cooling, refrigeration, and drinking water, using PV technologies. The high interest comes from many reasons, such as the use of domestic and free energy sources, increased energy security, and reduced dependency on fossil fuels and emissions. Earlier studies mainly focused on residential consumers. Recent studies have analyzed commercial scales such as school, hospital, and industry processes [10]. The key parameters or variables are plant type and configuration (e.g., solar with biomass, wind, geothermal, or hydrogen), energy inputs and outputs, and locations because renewable sources could vary from one region to another. The energy outputs are power, heating, cooling, and drinking water. Generally, most solar-based power generation technologies with steam Rankine cycle, organic Rankine cycle, Brayton cycle, small-capacity fuel cells, and storage systems are already commercialized. The storage options are electrical, hydrothermal, hydrogen, and compressed air.

The efficiency and reliability of solar-based technologies can be impacted by materials science, semiconductor physics, and optics. Solar-based power generation technologies are generally easy to use and are designed for long life with minimal maintenance. These technologies can be connected to the grid or used in stand-alone systems, and they can be designed to work with other renewable or conventional power generation systems. The existing technologies are relatively secure and can be designed to be resilient against cyber and physical threats. For example, solar PV technology has a range of efficiencies depending on the materials that can be impacted by shading or dust. Solar thermal systems have lower efficiency due to heat losses, and can be less affected by shading. Solar-based power and heat generation are among the most sustainable technologies with no or very low GHGs and environmental impacts during the operation. Producing solar panels (PV cells) has environmental impacts due to their unique materials that can be mitigated by recycling the used materials. Also, recycling the materials can reduce the total cost and make this technology more cost-competitive. Emissions from solar panel production are lower, and the costs have been decreasing, making this technology more reliable than other technologies. Reducing installation and maintenance costs can make solar panels more cost-competitive compared to other power generation systems. Currently, efficiency, durability, and cost reduction are the main parameters to improve the maturity of this technology. Countries in tropical or subtropical regions receive high amounts of solar radiation and can harness solar energy for power and heat generation most of the year. Small-scale applications of solar energy show social acceptability at both macro and micro levels, particularly in rural and remote areas that can use electrification for transportation, communication, and air conditioning. South-facing roofs are ideal for power generation using solar panels with at least five hours of sunlight, and west-facing roofs are the next best option. To summarize, each solar energy technology has its own advantages and disadvantages, and it highly depends on the technology type, location, and application. Solar-based power and heat generation provide a primary net-zero solution for addressing environmental degradation and climate mitigation.

1.1.1 Photovoltaic (Seven Technologies)

Solar PV technologies are the most researched approach for supplying both power and heat. PV panels can directly convert incident solar energy to power, following the principle of the photoelectric effect. Some of the most used materials in PV panels are crystalline silicon (C-Si), cadmium telluride (CdTe), copper indium gallium diselenide (CIGS), perovskites, multijunction (iii–v), and organic. PV array contains multiple solar cells, and several hundreds of solar arrays form a large-scale solar power generation device. The general performance and efficiency of PV systems highly depend on the semiconductor materials. PV systems can be either grid-connected or stand-alone, which are typically oriented to the south at north-facing latitudes and vice versa. Solar panels convert photons to electrons and generate around 200 terawatt-hours (TWh) of world power, and this figure could grow up to 11,000 TWh by 2050 and cut up to 70 gigatons of GHGs globally. Solar panels use silicon crystals to move subatomic particles and electrons to generate affordable electricity. Rooftop solar systems account for around 30% of worldwide capacity. Solar panels can generate up to 25% (approximately 17,000 TWh) of energy worldwide by 2050 from about 1% in 2020, and cut over 100 gigatons of GHGs. Cost reduction and incentives can accelerate rooftop solar system growth. Detailed comparisons of solar PV generation using different materials are provided by [8]. Figure 1.4 shows the LCOE progress and targets without state and local incentives.

Recent PV panels can follow the sun's path on a single or double axis to capture more solar radiation and increase their efficiency, but they require more area compared to fixed ones. The early applications were in satellites and spacecraft, and their recent applications have been in buildings, transportation, telecommunication, and rural electrification. The heat generated by PV panels can be used to run heat pumps in either heating or cooling mode, depending on the season. Earlier studies investigated the use of a hybrid energy system comprising solar PV panels for power generation and air-conditioning system, using the generated heat [12]. Also, it combined solar PV panels with a proton-exchange membrane fuel cell to generate power and heat for buildings in remote locations [13, 14]. The results

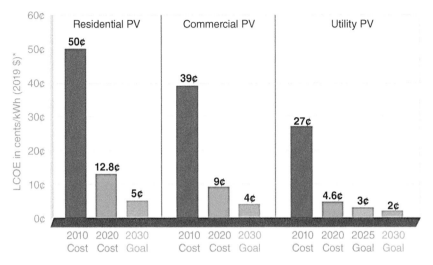

Figure 1.4 Solar PV progress and targets. Source: U.S. DOE [11]/Office of Energy Efficiency & Renewable Energy/Public Domain.

show that over half of the heat demands can be satisfied by excess hydrogen and waste heat from a fuel cell. Recent studies have explored mixed solar PV panels with thermal systems (PVT) that can be used for producing power, hydrogen, hot air, and drinking water with a PVT array as the primary conversion technology [15]. PVT array generates power and heat for various purposes, such as residential use or hydrogen production to run fuel cells and produce water or heat as outputs.

In 2020, the PV system price was $1.25 per watt for 25 years lifetime, and the annual installed capacity is about 19 gigawatts direct current (GWdc). The global projection is to reach 500 GWdc by 2030 to meet decarbonization scenarios [6]. For techno-economic analysis of the power plant, the main parameters are PV modules (37% of total cost), inverter (4–5% of total cost), electrical and structural balance of system (BOS) (21% of total cost), land, construction, engineering, and contingencies (20% of investment cost), operation and maintenance (2% of investment cost), interest rate, and insurance rate. PV mounting structures are mainly made of galvanized or stainless steel. The primary mounting structures are single-axis tracking ground mount, fixed-tilt ground mount, penetrating rooftop, and ballasted rooftop [6]. PV tracker systems can orient panels more directly toward the sun to increase power production per unit. PV inverters convert DC to AC, and are made of several components. The major cost drivers of inverters are semiconductor and electronic components, accounting for 48 and 30% of the cost, respectively. The inverter is one of the essential and expensive parts of PV power generation.

The conversion efficiencies of commercially available solar PV technologies are up to 22% [8]. The leading solar PV panels are made from C-Si and CdTe, which cover 84 and 16% of the U.S. market, respectively. According to DOE, the United States does not have active C-Si ingot, wafer, or solar cell production, and about 97% of the C-Si wafer production is in China, and about 75% of solar cells are made in southeast Asia countries, such as Malaysia, Vietnam, and Thailand. But the United States produces 16% of thin-file CdTe modules. Silicon, CdTe, CIGS, and hydrogenated amorphous silicon (a-Si:H) solar cells dominate 92, 5, 2, and around 1% of the PV market, respectively [16]. Flexible solar panels use PV materials (mainly thin film and C-Si) that are lighter, inexpensive, thinner, and easy to install. However, they are less efficient and less durable than standard solar panels. The 2022 average efficiency of PV panels is around 20% (370 W), highly depending on silicon type, cell design, and panel size.

Solar panels operate best at 25 °C (77 °F) with blue backsheets, and the initial cost is between \$15,000 and \$25,000. Solar PV systems are net-zero or low-carbon technologies due to their environmental impacts on land and water from the required hazardous and toxic materials (e.g., lead, chromium, cadmium, arsenic, mercury, and nickel) [17]. The GHG emissions from PV modules range from 14 to 73 g CO_2 eq. per kWh, which is 10 to 53 times lower than emissions from oil burning at around 742 g CO_2 eq. per kWh [17]. The benefits of flexible solar cells are: (i) suitability for curved (uneven) surfaces, (ii) cost-effectiveness compared to silicon solar cells due to fewer materials required, (iii) light weight, (iv) ease of installation and handling, and (v) reduced labor cost. Ultra-thin solar PV cells use different materials with excellent electrical and optical properties, such as transition metal dichalcogenides that can reduce material use, weight, and cost [18]. Also, they have several applications due to their longer battery life and flexibility to be molded into various shapes, such as sensors, wearable electronics, car roofs, or airplane wings.

1.1.1.1 Crystalline Silicon (TRL 10)

The principles and science behind C-Si solar panels are based on the physics of semiconductors and light properties. C-Si PV cells can be made of silicon wafers and treated with impurities to create a *p–n* junction (two different semiconductor types), allowing the cells to generate power from sunlight. The *p*-type semiconductor has positively charged holes, and the *n*-type semiconductor has negatively charged electrons. When the light photon is absorbed by the cells (silicon crystals), it can create electrical charge flows and can be used as a power source. The C-Si cell efficiency is measured by the amount of sunlight absorption and power generation. These cells have been widely used for power generation due to their high functionality and reliability to operate for several years without significant degradation. C-Si cells are easy to produce and can be used for various applications, such as stand-alone systems for powering remote sites or integrated into other structures (e.g., commercial or residential buildings). C-Si cells can be compatible with other systems and are generally considered to be a safe and secure method for power generation. Latest studies show the highest recorded efficiency of C-Si cells is around 27% due to their bandgap limitations since they can absorb a limited range of sunlight. They can be more efficient when they are exposed to direct sunlight.

Power generation via C-Si cells is a clean and renewable energy source that does not produce any GHGs or wastes during operation. However, producing C-Si cells has environmental impacts due to the use of different hazardous chemicals, which generate toxic wastes during the disposal process. Recycling these cells can address several sustainability challenges, such as waste management and material recovery. The main cost drivers for C-Si cell production are materials, manufacturing processes, and installation costs. Currently, over 95% of solar PV panels use C-Si, particularly made from polysilicon [19]. The top three main solar panels are monocrystalline, polycrystalline, and thin-film (Figure 1.5). The most efficient PV cells are monocrystalline, with 22% efficiency for up to 25 years, and they need less space compared to polycrystalline and thin-film. The main drawback of monocrystalline cells is the high production cost, which makes them less cost-effective for cold regions. Polycrystalline cells are the most effective and sustainable method. The main drawback is the low efficiency (up to 17%) due to high space requirements and the low silicon purity compared to monocrystalline

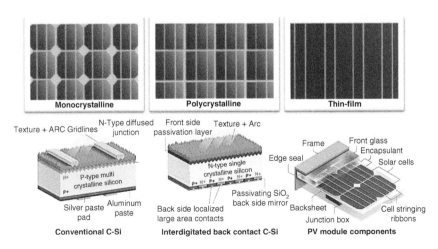

Figure 1.5 Schematic of C-Si solar cells. Source: U.S. Department of Energy, Solar Energy Technologies Office (public domain).

cells. Thin-film cells have several benefits due to their lower weight compared to other cells, which make them easy to move for remote applications. Their main drawback is low efficiency.

The major cost drivers for silicon cell production are silicon substrate (50%), module processing (30%), and device processing (20%) [16]. The bifacial cells absorb direct light from the sun and indirect light from the ground reflection that can increase power generation by 30%, particularly in deserts or lands without vegetation. In 2022, monocrystalline and thin-film PV cell costs were around $1–$1.5 per watt, and polycrystalline cell costs were around $0.9–$1 per watt. The cost of solar panel installation ranged between $15,000 and $25,000 [19]. Polysilicon production is the most energy-intensive and operational cost-driver step compared to the other steps in the solar supply chain. Currently, China is the largest producer of polysilicon, using coal-fired power. Wafer production wastes a third of silicon ingots that need recycling solutions to use the cost benefits of recycling.

According to the U.S. DOE and National Renewable Energy Laboratory (NREL), the production cost of C-Si PV is around $0.37/w in the United States and $0.28/w in China. The major cost drivers in the United States are materials, labor (22%), and overhead (profit) costs [6]. C-Si modules are made of high-purity polycrystalline silicon (polysilicon) from metallurgical-grade silicon (MGS) or silicon metals. The monocrystalline silicon ingots are made of melted polysilicon and sliced into thin silicon wafers for making solar cells and C-Si modules between glass and plastic. The required materials for C-Si PV are MGS, glass, encapsulant film, backsheets, and aluminum frames. Polysilicon is the primary PV material, and the main polysilicon production method from MGS is Siemens chemical vapor deposition. Over 90% of the market uses the Siemens method, and the rest applies the fluidized bed reactor method, covering around 5% of the market [6]. Polysilicon production has high capital and operational costs for building a plant and refinery process with high skilled labor and high power demand. Currently, China produces 72% of polysilicon globally, and the price increased from $6/kg in 2020 to $28/kg in 2021. After China, Germany and the United States have the largest polysilicon production capacity. The main difference between the U.S. and Germany polysilicon compared with China is the semiconductor quality, which is 11N (11 nines or 99.999999999%) or greater.

MGS production process is used for making polysilicon that can be used in solar wafers and semiconductors, using quartz (silica dioxide) and energy-intensive furnace (10–15 MWh for a ton of MGS). According to the U.S. Geological Survey (USGS), China currently produces over two million tons of MGS (around 70% globally). The primary material for the solar-grade front glass is silica sand with low iron for optimal transmissivity of sunlight, typically 3.2 mm glass. The rear glass for bifacial C-Si or thin-film panels is a 2-mm soda–lime glass.

China is the largest producer of PV coverglass. Encapsulant film forms barriers around PV cells that are made of ethylene vinyl acetate for monofacial PV panels or polyolefin elastomers for bifacial or thin-film panels. Backsheets are made of PET (core layer) and PVF or PVDF (out layers). Regular backsheets are mainly used in monofacial C-Si panels, and clear backsheets are used in bifacial panels to protect them from moisture and wind. According to the DOE reports, 50% of DuPont's PVF goes to PV backsheet applications [6]. The primary processes for PV wafer production from polysilicon are Czochralski process for monocrystalline wafers and the directional solidification process for multicrystalline wafers, using high temperatures (over 1,400 °C). It takes four days to produce Czochralski ingot at a 1-mm/min growth rate and three days to produce multicrystalline ingots. The produced ingots are sliced into thin wafers (around 180 μm thick), and a third of the ingot is wasted during the process. Solar cell production requires several chemical processes for converting wafers to cells, including high-temperature diffusions, coating depositions, and metallization. Due to high efficiency, the passivated emitter and rear cell structure currently dominate the market. Silver is another primary material in C-Si cells, accounting for about 10% of cell cost, and the PV industry used around 10% of global silver in 2019.

In summary, C-Si solar cells are a widely used and affordable technology for generating power from sunlight. They are reliable, easy to manufacture, and can be used in various applications. The cells are compatible with a wide range of equipment and systems, and they are considered to be a safe and secure technology. From a sustainability standpoint, silicon solar cells are a clean and renewable energy source, but their production and disposal can have environmental impacts. The cost of silicon solar cells has decreased significantly over time, making them more affordable and competitive with other sources of electricity. An overview of C-Si bifacial PV panels and power plants in different locations has been provided by [20].

1.1.1.2 Multijunction (TRL 9)

The principles and science of multijunction cells are similar to those of C-Si cells, which are based on light properties and semiconductor physics. The main difference is the multiple layers of semiconductor materials in multijunction cells with several bandgaps for absorbing more solar spectrum. Multijunction cells are more efficient than C-Si cells, making them ideal for applications requiring a high power-to-weight ratio, such as unmanned aerial vehicles and satellites. Multijunction cells are compatible with various equipment and systems and have the highest recorded efficiency up to 47%, especially for concentrated solar power (CSP) systems with direct sunlight. Power generation with multijunction cells is a clean

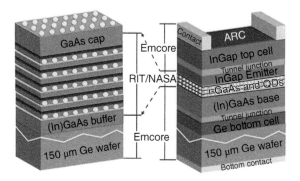

Figure 1.6 Schematic and example of a multijunction solar cell. Source: Kerestes et al. [23]/John Wiley & Sons.

and renewable method that does not release any GHG emissions or waste during operation; however, the production and disposal of multijunction cells can have environmental impacts due to the use of hazardous chemicals. Recycling multijunction cells is becoming increasingly common, which can reduce waste and recover valuable materials. The cost of multijunction solar cells is higher than that of C-Si solar cells due to the complexity of their manufacturing process and the use of more expensive materials.

The photoelectric conversion efficiency of single-junction solar cells is low (around 20%) because they can absorb part of the solar spectrum. To absorb multi-band light of solar radiation, multijunction PV cells have been developed with a solar cell stacked with semiconductors at different junctions (bandgaps) that can absorb photons in other regions with record power conversion efficiency of up to 45% [21, 22]. Earlier studies investigated the number of junctions and semiconductor properties of multijunction panels, particularly elements in columns III and V in the periodic table, such as gallium indium phosphate and gallium indium arsenide (Figure 1.6). The results showed that the efficiency of multijunction PV cells increases with the number of junctions. Also, three-junction cells from III–V semiconductors have over 45% efficiency, utilizing concentrated light. Concentrating optics and a dual-axis sun-tracking system can increase the efficiency of multijunction III–V cells and help address the high cost of their semiconductor substrates. The main benefits of multijunction PV cells are: (i) spectrum matching with specific absorber layers and (ii) crystal structure and properties, such as compatible absorption spectra.

In summary, multijunction PV cells are highly efficient technologies that are ideal for specialized applications, such as CSP systems and space applications. They can generate clean and renewable power, but their cell production and disposal can have environmental impacts, and their cost is higher compared to C-Si PV cells.

1.1.1.3 Floating Solar PV (TRL 8)

Floating solar PV systems use panels that can be placed on floating platforms for power generation from sunlight (Figure 1.7). The concept is similar to land-based solar PV systems, and the main differences include the water environment, such as buoyancy, wave and wind loads, and water quality. This technology can be used in

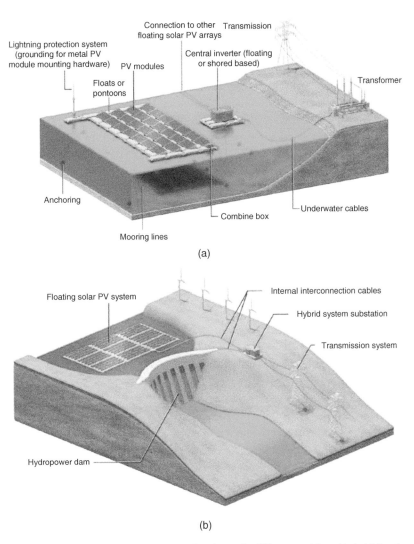

Figure 1.7 Schematic of stand-alone floating solar PV system (a) and hybrid floating solar PV with hydropower system (b). Source: Gadzanku et al. [24].

areas where land-based solar installations are limited (e.g., reservoirs, dams, and other water bodies). It can reduce water evaporation and increase water quality by limiting the amount of sunlight that reaches the water surface. Also, floating solar PV systems can be used for both large-scale power generation and off-grid applications, and can be compatible with other equipment and systems. This technology is safe and secure, similar to land-based solar PV systems, and their main benefits compared to land-based solar PV systems include higher efficiency due to the cooling effect of water and fewer land use conflicts. However, their performance can be affected by several factors, such as wind, waves, water quality, and platform design.

Floating solar PV systems can generate clean and renewable energy without releasing any GHG emissions or waste during operation. They can also help reduce water evaporation and increase water quality. However, the production and disposal of floating solar PV systems can have environmental impacts, and their anchoring systems can potentially cause damage to the underwater environment. Other drawbacks are algae bloom growth and other depositions (e.g., salt) because they can degrade the PV systems over time and negatively affect the ecosystem. Earlier techno-economic studies show that the cost of floating solar PV systems can be higher by 30% compared to land-based solar PV systems due to the additional costs of the floating platform, anchoring systems, and other water-specific requirements. This technology is expected to grow and address land-based solar systems' challenges and limitations. More detailed information about floating solar PV systems is provided by [24, 25].

1.1.1.4 Cadmium Telluride (TRL 6)

CdTe solar cells use a thin-film PV technology, incorporating CdTe as the light-absorbing material. The science behind CdTe PV cells is based on the photoelectric effect, where photons from sunlight are absorbed by CdTe layer, generating electron-hole pairs separated by the electric field within the cell to produce power. CdTe cells are commonly used in large-scale utility projects due to their high efficiency and low cost compared to other PV technologies. They are also used in small-scale residential and commercial applications. CdTe cells can be manufactured in thin, flexible sheets that are easy to install, making them a popular choice for buildings. CdTe cells are compatible with various equipment and systems, but their use may require special considerations due to the presence of cadmium, a toxic heavy metal. However, their performance can be affected by several factors, such as temperature, shading, and the quality of the substrate material. CdTe cells have a low carbon footprint, as they require less energy to manufacture and generate power without releasing GHG emissions during operation. However, the production and disposal of CdTe cells can have environmental impacts due to the use of cadmium. Proper handling, disposal, and recycling of CdTe cells are necessary to minimize the environmental impacts. However, their production may require specialized equipment and processes that can increase initial costs.

According to the U.S. DOE, CdTe cells are the second most common PV after C-Si due to high absorption with around 5% market share [26]. CdTe cells can be produced inexpensively with a record efficiency of 22.1% in the lab and 18% at the commercial level. Among thin-film PV technologies, CdTe-based PV panels lead the market with an annual production of 2 GW [8]. The United States is the largest producer of thin-film CdTe PV modules; however, the main challenge is the required materials because Cd is toxic and Te is a rare element. The main CdTe thin-film deposition methods are vapor-transport and close-spaced. The production cost of CdTe PV is around $0.31/watt in the United States without considering shipping costs, and in southeast Asia with considering the shipping cost to the United States. Cd and Te are both byproducts of other metals, such as gold, copper, and zinc. Over 80% of Cd is produced from smelting zinc ores, and over 90% of Te is made mainly from copper refinery processes. Currently, the largest use of Te is in PV panels. High-purity Cd and Te are required that can be achieved through electrolytic purification followed by atomization or via vacuum distillation. Their major cost driver is the required materials. CdTe solar cells have high TRL and have been in commercial use for several decades.

The latest studies focused on making CdTe solar cells cheaper and more efficient by improving crystal quality, doping control, and lifetime, as well as material reuse and recycling to address material scarcity and hazard concerns. Particularly, researchers explore new CdTe contacting materials to absorb light from the front and back of the bifacial CdTe modules. For example, a French company developed an ultra-thin PV cell (called ASCA) for recharging a car battery that can be used as a car cover and rolled/unrolled 1,000 times without issues or performance loss. It can charge the vehicle battery up to 9 miles (15 km) from eight hours of sunlight. This technology can power water pumps, satellites, or encampments [27]. Overall, CdTe cells are a highly efficient and low-cost PV technology suitable for large- and small-scale solar power generation. However, the presence of cadmium in the cells requires proper handling, disposal, and recycling to avoid environmental contamination.

1.1.1.5 Copper Indium Gallium Diselenide (TRL 6)

CIGS solar cells use a thin-film PV technology that incorporates copper, indium, gallium, and selenium as light-absorbing materials. The science behind CIGS PV cell is based on the photoelectric effect, where photons from the sunlight are absorbed by the CIGS layer, generating electron-hole pairs that are separated by

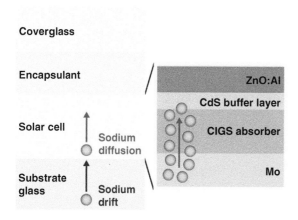

Figure 1.8 Schematic of the CIGS PV cell. Source: Adapted from Kettle et al. [28].

the electric field within the cell to generate power (Figure 1.8). CIGS cells are suitable for both large- and small-scale solar power generation, and they can be manufactured in flexible, lightweight sheets that are easy to move and install. CIGS cells are also suitable for buildings due to their ability to be incorporated into building materials. CIGS cells are compatible with various equipment and systems and are safe to use. They do not need toxic materials (e.g., cadmium) that are in other thin-film PV technologies.

The main benefits of CIGS solar cells are: (i) high efficiency, which can achieve conversion efficiencies of up to 22%, (ii) good low-light performance, (iii) fewer problems and less susceptibility to shading, and (iv) a relatively low carbon footprint since they need less energy to manufacture for power generation compared to other PV technologies. CIGS cells are more sustainable compared to other cells due to the use of nontoxic materials in their production and their ability to be incorporated into building materials. Earlier studies show CIGS cells have relatively lower manufacturing costs, but their production may require unique processes that can increase the total costs. CIGS solar technology has TRL 6, and has been in use for several years. The technology is constantly improving, with ongoing research focused on increasing efficiency and reducing manufacturing costs. CIGS solar cells do not release GHG emissions during operation, and their production and disposal have relatively low environmental impacts compared to other PV technologies.

According to the U.S. DOE, CIGS thin-film cells showed high efficiency, around 20% in the lab and approximately 14% in commercial panels, due to their high absorption, tandem design, and protective buffer layer [29]. The most popular and cost-effective deposition methods to form CIGS films are co-evaporation in high-vacuum environments and two-stage deposition using precursor reaction processes such as electroplating and sputtering. There are other methods, such as three-stage deposition, reactive or magnetron sputtering, and electrodeposition. Recent studies investigated various glasses (e.g., soda–lime) and the role of sodium (Na) in increasing CIGS performance since it is inexpensive compared with other glass substrates [29]. In addition, the role of molybdenum (Mo) has been explored due to its high reflectivity as the back contact that can be deposited through DC sputtering process. Latest studies investigated flexible thin-film solar cells from CdTe and CIGS, as well as antimony selenide (Sb_2Se_3) and a-Si:H due to fast and easy installation, less waste, and flexibility to adapt various shapes [16, 30]. CIGS cells' efficiency and stability are comparable to those of Si cells, but the main barriers are the high-temperature processing of CIGS and the high production cost.

In summary, CIGS cells are a highly efficient, low-cost, and sustainable PV technology that can be suitable for both large- and small-scale solar power generation. Their use of nontoxic materials and ability to be incorporated into building materials makes them a safe and versatile option for solar power generation.

1.1.1.6 Organic (TRL 6)
Organic solar cells use thin-film PV technology to incorporate organic molecules as light-absorbing materials. Similar to CIGS solar cells, the science behind organic cells is based on the photoelectric effect, where photons from the sunlight are absorbed by organic materials, generating electron-hole pairs that are separated by the electric field within the cell for power generation (Figure 1.9).

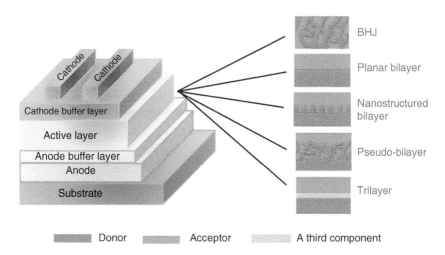

Figure 1.9 Schematic of the organic solar cell. Source: Wang et al. [31]/John Wiley & Sons.

Organic PV cells are suitable for small-scale solar power generation and can be manufactured in flexible, lightweight sheets that are easy to move and install. Organic cells are also suitable for buildings due to their ability to be incorporated into building materials. Organic cells are compatible with various systems and are safe to use. Like CIGS cells, organic cells do not contain any toxic materials (e.g., cadmium or lead).

This technology has lower efficiency compared to other PV technologies, with the highest conversion efficiencies currently around 17%. Other drawbacks are lower durability and a shorter lifespan that can sometimes limit their practical application. Organic solar cells are more sustainable due to: (i) their low carbon footprint and less energy use for manufacturing, (ii) their potential to be produced from renewable materials and have lower environmental impacts, and (iii) their potential to be low cost, using printing techniques that use fewer materials. However, their low efficiency limits their cost-effectiveness compared to other PV technologies. Organic solar cells do not release GHG emissions during operation, and their production and disposal have relatively low environmental impacts compared to other PV technologies.

Organic PV cells are made of organic (carbon-rich) materials and use molecular or polymeric absorbers with various features, such as transparency. However, their efficiency is around 11% (half of C-Si cells), and they have shorter lifetime compared to inorganic ones [32]. Organic PV cells are classified to: (i) small-molecule cells with high absorption capacity in visible and near-infrared spectrum and (ii) polymer-based cells with high interface surface area. Their low efficiencies are due to low carrier mobilities and small diffusion lengths that can be addressed by improving absorber materials [33]. Organic PVs can be used in various shapes and materials, such as plastics, and require a low-cost manufacturing approach on a large scale. The benefits of organic PVs are: (i) inexpensive production, using roll-to-roll processes, (ii) abundant materials, and (iii) flexibility that can allow various applications. Some companies (e.g., Epishine in Sweden [34]) use organic electronics from hydrocarbon chains or polymers without using silicon or metals for producing printable organic PV cells. Overall, organic PVs are a promising technology with the potential to be a low-cost and sustainable method that is suitable for small-scale solar power generation and buildings. However, their current low efficiency and durability limit their practical application in some cases, and further research is needed to improve their performance and address degradation in harsh climates and industry-relevant lifetime.

1.1.1.7   Perovskite (TRL 4)

Like CIGS and organic cells, perovskite solar cells are a type of thin-film PV cell that uses hybrid organic–inorganic materials with a perovskite crystal structure as

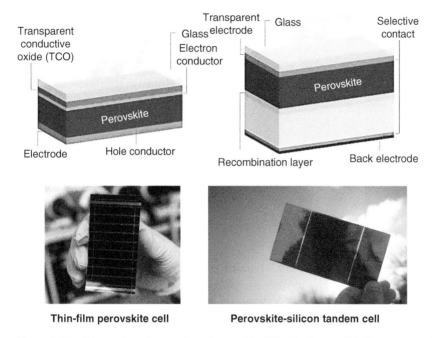

Thin-film perovskite cell   **Perovskite-silicon tandem cell**

Figure 1.10 Schematic and examples of perovskite PV cells. Source: U.S. Department of Energy, Solar Energy Technologies Office (public domain).

the light-absorbing layer (Figure 1.10). The perovskite material is relatively new in PV technology, but it has shown high power conversion efficiency and has the potential for low-cost production. Perovskite cells are lightweight, flexible, and can be manufactured in various sizes and shapes. They are suitable for large-scale solar farms and small-scale applications, such as portable devices or buildings. Perovskite cells can be used with other systems since they are safe and do not contain toxic materials. Recent studies show that perovskite cells' efficiency can reach 25%. They also have the potential for high performance in low-light conditions that can benefit indoor applications. The main drawbacks are: (i) limited durability and stability over time and (ii) they may be susceptible to degradation due to moisture and other environmental impacts.

Similar to other thin-film PV cells, perovskite cells have low environmental impacts and costs due to less energy used for producing them. Also, they do not release GHG emissions during operation, and their production and disposal have relatively low environmental impacts compared to traditional solar cells (e.g., C-Si). However, their current challenges (e.g., durability and stability issues) and the environmental impacts of large-scale production and disposal of perovskite cells need to be addressed to improve their effectiveness in the long term. Perovskite cell technology has TRL 4 and is in the early research and development stages.

Perovskite is a class of crystalline structures that showed effective sunlight conversion to power and effective power conversion efficiency of over 25%; however, they are not commercially viable because of their limited stability and efficiency reduction over time [35]. After multijunction III–V PV, perovskite PV cells exceeded power conversion efficiency on the laboratory scale. Perovskite materials can decompose over time, reacting to light, heat, oxygen, or moisture. The latest studies reported that perovskite efficiency is around 15% over a month of outdoor production. Perovskite PVs with high performance levels contain water-soluble lead, which requires adequate plans for lead replacement after usage. The major production types are sheet-to-sheet (rigid base) and roll-to-roll (flexible base).

The U.S. DOE and NREL studies show that the efficiency records of single-junction perovskite and tandem perovskite–silicon are 25.7% for 29.8%, respectively [35]. Currently, two companies (Oxford PV in Germany and GCL in China) claim to use perovskite cells and will produce up to 100 MW in 2022 [6, 36]. The main issue with flexible perovskite panels is the moisture-induced degradation compared to flexible CdTe, CIGS, and a-Si:H panels. GaAs two junction (non-concentrator) and tandem perovskite–silicon PV cells showed the highest efficiency records, which are 32.9 and 29.8%, respectively [35].

1.1.2 Solar Thermal Power (Four Technologies)

Solar thermal power technology, also known as CSP, is based on several established scientific principles, such as light-to-heat conversion and heat transfer to working fluids to drive a turbine and generate power. Solar thermal power systems can provide baseload electricity and generate power continuously and reliably, similar to traditional power plants. They can also be integrated with other equipment and systems, such as renewable energy sources, to provide a more stable and reliable power supply. Solar thermal power plants require significant amounts of land, sunlight, and water, along with large-scale installations. They can be used in various settings, but are most effective in areas with abundant sunshine and available water resources.

Solar thermal power systems can provide reliable, dispatchable power to help balance fluctuations in other renewable energy sources. Solar thermal power plants are generally considered secure and reliable, but like any large-scale infrastructure, they are vulnerable to natural disasters and other security threats (e.g., cyberattacks). However, the risks can be mitigated through careful design, operation, and maintenance. Solar thermal power plants can achieve high levels of efficiency, with some designs reaching thermal-to-power conversion efficiencies of over 40%. However, the performance of a solar thermal power plant is highly dependent on the availability of sunlight and water, as well as the design and operation of the plant.

Solar thermal power is a renewable energy source without releasing GHG emissions or other pollutants during operation. However, the construction and operation of solar thermal power plants can have environmental impacts, particularly regarding land use and water consumption. The cost of solar thermal power systems has decreased in recent years, but it remains higher than many other forms of renewable energy, such as solar PV. Solar thermal power systems can be cost-competitive in specific regions and applications, particularly in areas with high power prices or limited access to other energy sources. Improving the efficiency of these processes and developing new materials and technologies can help reduce costs. This technology is mature and has high TRL. There are many commercial-scale solar thermal power plants in operation around the world.

Solar thermal power systems utilize mirrors (reflectors) and lenses to concentrate light beams from a large area to a small area by converting the heat to power through thermodynamic cycles. They can generate power in the absence of sun with high efficiency, but they require high capital costs and are not suitable for small-scale solar plants. Solar thermal collectors can be used for combined heat and power (CHP) applications, such as power generation and hot sanitary water from solar heat in the Mediterranean region or California in the United States [37]. These solar thermal systems use different light concentrators that depend on thermodynamic efficiencies of sun-tracking and insolation focus for generating heat. Recent innovations in light concentration (e.g., small heat loss) have improved CSP systems and reduced the total costs. The investment cost of a CSP plant is similar to that of a PV power plant, and the differences are in the parabolic trough, power block, and thermal storage costs instead of PV modules, inverter, and BOS in PV plants. Figure 1.11 shows the LCOE progress and targets without state and local incentives.

Solar thermal systems or CSP technologies showed higher annual power generation than PV panels. However, PV systems showed better yields and performance, considering the same land use for power generation, because CSP systems require high land use [38]. The calculated GHG emissions of parabolic trough and tower CSP systems are around $23\,g\,CO_2$ eq. per kWh [39, 40], and around $50\,g\,CO_2$ eq. per kWh for the thin-film PV technologies. The environmental impacts of using solar PV panels are higher than those of CSP systems during the whole life cycle due to the required hazardous, toxic materials for manufacturing PV cells and cleaning the semiconductor surface, such as silicon dust [41]. Solar power generation is a net-zero solution; however, there are emissions from other solar life cycle phases, such as material manufacturing and transportation, maintenance and cleaning,

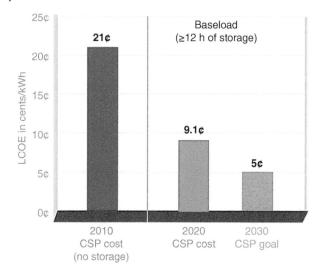

Figure 1.11 CSP progress and goals. Source: U.S. Department of Energy, Solar Energy Technologies Office (public domain).

and decommissioning phases [42]. The maintenance costs of the PV power plant and CSP plant are 1 and 2% of the initial investment cost, respectively [8].

Local, national, and global supports and legislative regulations can help solar CHP plants compete with fossil fuel-based plants and promote net-zero or low-carbon technologies to meet energy needs and ambitious climate targets. Fully solar-based CHP plants can be used to generate local energy needs using free local sources. Overall, fully renewable solar CHP plants are more economically and environmentally sustainable than fossil fuel-based plants [10]. However, solar-based energy, as the only input without any storage capacity, cannot satisfy power and heat demands due to consumer needs and solar availability.

1.1.2.1 Solar Thermal Tower (TRL 9)

Solar thermal towers are a type of CSP technology that uses mirrors or lenses to focus sunlight onto a central receiver that can be converted into thermal energy and used to generate power through a heat engine or steam turbine. This technology is suitable for large-scale, utility-scale power generation, as well as small-scale applications, such as remote communities or industrial processes. Solar thermal towers are particularly effective in areas with high solar irradiance (power per unit area received from the sun's radiation). Solar thermal towers can be integrated into existing power systems and used in conjunction with other renewable or non-renewable energy sources.

The main benefits of solar thermal towers include: (i) high thermal-to-power conversion efficiencies (30–40%), higher than many other types of renewable energy; (ii) energy storage that allows them to provide power even when the sun is not shining and makes them more reliable and dispatchable compared to some other renewable energy sources; (iii) scalability, which makes them suitable for a wide range of applications; (iv) low operating costs that can operate for many years with minimal maintenance; (v) no GHG emissions or other pollutants during operation, making them a clean energy source; and (vi) operating at high temperatures that can be used in industrial processes with high-temperature heat requirements. The main drawbacks are (i) high land use to install the heliostats and tower that can be a challenge in areas where land is limited or expensive; (ii) high water consumption to generate steam and to cool the receiver that can be a significant constraint in areas with limited water resources; (iii) high capital costs than many other types of renewable energy, which can be a barrier to their deployment; (iv) technical complexity and a high level of expertise to design, build, and operate, which can add to the costs and risks associated with the technology; and (v) weather dependency, which can be impacted by weather patterns and seasonal changes and can affect their performance and reliability, and may require additional backup power sources.

Solar towers use a large number of heliostats to concentrate sun rays to a central receiver, as many as 1,500 times, and heated fluids (e.g., molten salt) to generate power from steam (Figure 1.12) [44]. Their capacity varies between 10 and 200 MW, and their current TRL is around 9, meaning it is commercially viable, but needs evolutionary improvement to stay competitive. Solar towers require a storage system, and the solar-to-power efficiency is around 7–20% annually. Their temperatures are between 250 and 560 °C. Currently, two solar power tower facilities are operating in the United States, which are Ivanpah Solar Power with 393 MW in California and Crescent Dunes Solar Energy with 110 MW in Nevada.

1.1.2.2 Parabolic Trough (TRL 9)

Parabolic trough solar thermal power systems consist of a linear parabolic reflector to concentrate sunlight onto a receiver tube along the reflector's focal line (Figure 1.13). The receiver tube contains a heat transfer fluid that can absorb the concentrated solar energy and transfer it to a heat exchanger for power generation or providing heat for industrial processes. Parabolic trough systems can generate power and heat in remote areas or generate distributed power in urban or industrial settings. This technology has several applications, such as powering desalination plants and air-conditioning systems, and can be integrated into existing power systems and used with other renewable or non-renewable energy sources.

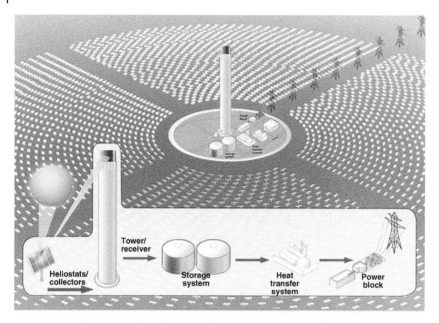

Figure 1.12 Schematic of a solar thermal tower receiver with heliostats field. Source: Ho [43]/John Wiley & Sons.

The benefits of parabolic trough systems include: (i) achieving high thermal efficiencies due to the concentration of sunlight in a small receiver area, (ii) operating at high temperatures, suitable for industrial processes with high-temperature heat requirements, and (iii) being a sustainable energy source as they do not release GHG emissions or consume fossil fuels during operation, and their overall environmental impact is relatively low due to the manufacturing and installation of the components required for these systems. The main drawbacks are: (i) the high cost of parabolic trough systems, (ii) receiver tube degradation, and (iii) oil-based heat transfer media that can limit output to moderate steam.

Parabolic trough is one of the major types of linear concentrator systems where the receiver tubes are located along the focal line of each mirror [44]. Several studies investigated a modular CHP plant, using parabolic trough collectors for generating power and thermal power in small-scale or stand-alone plants in isolated regions [46, 47]. The total cost of power generation was found to be competitive with lower GHG emissions. Recent studies investigated a coupled organic Rankine cycle with a parabolic trough solar thermal power system [48]. The results indicated that the overall plant could increase efficiency by up to 30%. Parabolic troughs can use a single axis or dual axis to track the sun, where the parabolic sheet is heated and generates power. Their capacities are 10–300 MW,

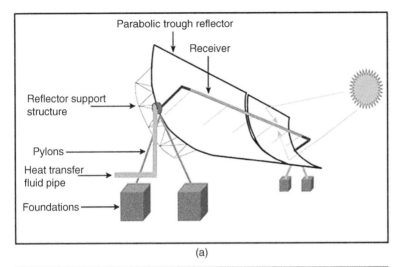

(a)

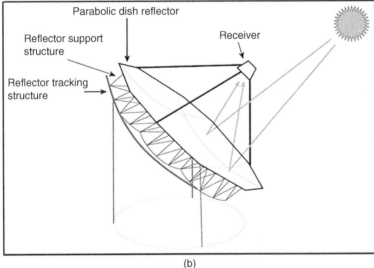

(b)

Figure 1.13 Schematic of the parabolic trough (a) and parabolic dish (b). Source: Abed [45]/John Wiley & Sons.

and their TRL is around 9, similar to solar thermal towers and parabolic dishes. Their temperature is between 350 and 550 °C, their solar-to-power efficiency is between 11 and 16% annually, and they need storage systems. Parabolic troughs are the most mature CSP technology that can produce heat at high temperatures.

Currently, parabolic trough collectors dominate the global market and account for 90% of solar thermal plants, and several parabolic trough solar thermal electric facilities are operating worldwide and in the United States, such as Solana Generating Station with 296 MW capacity in Arizona, Mojave Solar Project with 275 MW capacity in California, and Genesis Solar Energy Project with 250 MW capacity in California [44]. In summary, parabolic trough solar thermal power systems are a promising technology for large-scale, sustainable power generation and high-temperature heat applications. They are reliable and can be used for various applications, but the high cost and environmental impacts of their components need to be addressed to increase their deployment and competitiveness with other renewable energy technologies.

1.1.2.3 Parabolic Dish (TRL 9)

Parabolic dish solar thermal power systems are a type of CSP technology using parabolic-shaped mirrors to focus sunlight onto a small receiver at the dish's focal point (Figure 1.13). The receiver absorbs the concentrated solar energy and converts it to heat for generating power or heat. Parabolic dishes (dish-Stirling) use mirrored dishes to concentrate sunlight onto a central thermal receiver, transfer the heat to the engine generator, and convert it to power. This technology can handle high temperatures between 550 and 750 °C, and it does not need storage systems. The generated power can be distributed to the main grid or used in remote areas for different applications.

Similar to other solar thermal systems, the benefits of parabolic dishes include: (i) achieving high thermal efficiencies, (ii) operating at high temperatures, and (iii) being a sustainable energy source, as they do not release carbon emissions during the operation. Like other solar thermal systems, this technology requires high capital costs compared to other net-zero technologies, such as solar PV cells. Parabolic dishes are not well-suited for large-scale power generation applications, and there are no commercial solar dish operations in the United States [44] due to high capital costs, technical complexity, weather dependency, and land use. The TRL of parabolic dish systems is around 9, and several large-scale systems are already in operation worldwide with a capacity of around 12–25 MW. This technology has high potential in the near future, particularly for smaller-scale applications. Further research is needed to improve efficiency, reduce costs, and increase deployment.

1.1.2.4 Linear Fresnel Reflector (TRL 8)

Linear Fresnel reflector (LFR) is another type of CSP technology that uses sunlight to generate heat and power. The basic scientific principle behind LFR technology is the conversion of solar radiation into heat energy for generating steam to drive a turbine and generate power (Figure 1.14). In LFR systems,

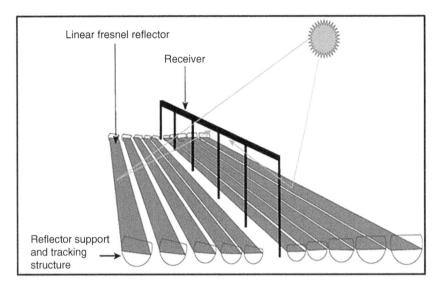

Figure 1.14 Schematic of the linear Fresnel reflector. Source: Abed [45]/John Wiley & Sons.

sunlight is focused onto a linear receiver using a series of flat mirrors or reflectors. LFR technology uses mirrors or reflectors to concentrate sunlight onto a linear receiver, usually a tube filled with a heat transfer fluid. LFR is similar to parabolic trough systems, using a linear concentrator system where a receiver tube is located above several mirrors and uses Fresnel lens for tracking the sun [44, 49].

LFRs use long and thin mirror segments to concentrate incoming rays (up to 30 times their average intensity) onto fixed absorbers that generate heat and use a heat exchanger to run steam generators. Their capacity is between 10 and 200 MW. The solar-to-power efficiency is around 13% at 390 °C and they need a storage system. The main benefits of LFR systems are high concentrations of sunlight, modular design, low water usage, and lower cost in comparison to other CHP systems, such as parabolic trough collectors. The main drawbacks are low efficiency, land use, and complex storage integration.

LFR systems can be used in both small- and large-scale applications in various settings, including utility-scale power generation, industrial process heat, and desalination. This technology can be used in hot and cold climates, although it is more effective in areas with high solar radiation. LFR technology is compatible with different thermal energy storage systems for stabilizing power supply during periods of low solar radiation. LFR systems are secure and do not use hazardous materials or processes. But they can be vulnerable to weather-related or other physical damage. The LFR performance depends on several factors, such as the quality of the mirrors or reflectors, the design of the receiver, and the efficiency of the power conversion system. LFR systems have lower thermal efficiency compared to other types of CSP systems.

LFR technology is a renewable and sustainable method for power and heat generation, using sunlight without emitting emissions or pollutants. The production and disposal of LFR components may have environmental impacts. Earlier techno-economic studies show that the main cost factors of LFR systems include the size and capacity of the system, materials, and installation and maintenance costs. LFR technology is considered to be at a mature stage of development, with various commercial systems already in operation around the world. Their TRL is 8, meaning they are ready for commercial demonstration and full-scale deployment in final form. With ongoing research and development, these systems have the potential to become an increasingly important source of clean energy in the future.

1.2 Wind

Wind is a renewable and abundant energy source that can be used in numerous ways, such as running old windmills on farms to pump water for agricultural purposes or new wind turbines to generate power for residential or commercial buildings [50]. The main principles of wind energy are based on wind patterns and speeds (shear) that can change by the earth's rotation, surface irregularities, and the uneven warming of the atmosphere by the sun [51]; for instance, a hot place near tall mountains is suitable for a lot of wind. Table 1.3 presents some of the top countries with high wind power generation by the end of 2021.

Wind power generation relies on the natural movement of air molecules in the atmosphere, which creates a force that can be harnessed by wind turbines. The principle of wind turbines is very similar to that of windmills, which capture wind energy. Wind turbines use aerodynamic force from the rotor blades to spin the generator and produce power, falling into two main types: the horizontal or vertical axis. Horizontal-axis turbines are the most common ones, with a tower (average 91 m tall) and three blades (average 61 m long), facing into the wind at the top of the tower, using a weather vane to keep the turbine turns into the wind and captures the most energy. Vertical-axis turbines have several varieties and do not need to be adjusted to the wind [51]. Wind turbine is a net-zero, low-carbon technology for power generation and can be installed on land or in oceans or seas to capture powerful wind and generate from 100 kW to 1 MW of power. Power generation from wind can be classified into three types: onshore, offshore, and airborne (Figure 1.15). Distributed wind turbines have applications in hybrid energy systems (e.g., solar) for residential, agricultural, and industrial applications. Earlier

Table 1.3 Top 10 countries with high wind power capacity.

Annual capacity (GW)		Cumulative capacity (GW)	
China	47.57	China	338.31
United States	13.41	United States	135.89
Brazil	3.83	Germany	64.54
Vietnam	3.49	India	40.08
United Kingdom	2.64	Spain	28.32
Sweden	2.10	United Kingdom	26.58
Germany	1.92	Brazil	21.58
Australia	1.74	France	19.13
India	1.46	Canada	14.25
Turkey	1.40	Sweden	12.10

Source: Adapted from US DOE [52].

techno-economic studies show that the price of power generation with wind turbines was between $800 and $950 per kW, and the average installation cost was $1,500 per kW in 2021, over 40% from a decade ago [52]. The utility-scale wind power costs around $32 per MWh.

The latest U.S. DOE studies reported that wind turbines are becoming taller to capture more energy since wind increases at higher heights and altitudes [53]. The average hub height for land-based and offshore wind turbines was 98 m (308 ft) and 150 m (330 ft) in 2021, respectively. Offshore wind speeds are at the highest rate during the demand peak in the afternoon and evening, representing a major opportunity near coastal cities. Offshore wind turbines can generate over 2,000 TWh from 60 TWh, and cut over 11 gigatons of GHGs by 2050. Onshore

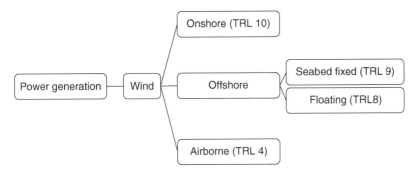

Figure 1.15 Leading wind technologies for power generation.

Table 1.4 Emission (CO_2, SO_2, NO_x) benefits by the U.S. regions in 2021.

Region	Wind (MWh)	Total benefits ($/MWh)
California	15,794,202	47
Central	82,358,257	83
Mid-Atlantic	22,382,067	102
Midwest	73,543,713	125
New England	3,547,754	28
New York	4,391,524	32
Northwest	41,392,012	46
Southwest	2,977,195	71
Texas	87,106,449	62
National	333,493,173	80

Note: Estimates were not provided for the southeast due to the small number of wind plants in that region.
Source: Adapted from US DOE [52].

(land-based) wind turbines can generate over 25% of energy from around 4%, and cut over 140 gigatons of GHGs by 2050 (Table 1.4). Texas, Oklahoma, New Mexico, and Kansas installed the most wind capacity in 2021.

In the United States, the wind industry has 136,000 MW capacity (1.5–4 cents/kWh), employs over 120,000 employees, and produces most of the components (over 8,000 parts), led by General Electric, Vestas, Siemens-Gamesa Renewable Energy, and Nordex with 47, 26, 13, and 13% of the U.S. market, respectively. In the United States, 41 states have utility-scale wind power capacity, and 50 states have distributed wind power systems installed, which supplies 20% of total power in 11 states and over 95% total U.S. power generation in 2021 [54]. Total wind power generation capacity reached 136 GW in 2021 (Table 1.5), ranking the United States as the second largest source of wind power with 32% of global capacity after China

Table 1.5 Annual and cumulative growth in U.S. wind power capacity over the last 10 years.

	2011	2012	2013	2014	2015	2016	2017	2018	2019	2020	2021
Annual (GW)	6.6	13.3	1.1	4.9	8.6	8.2	7.0	7.6	9.1	17.2	13.4
Cumulative (GW)	46.8	60.1	61.1	65.9	73.9	82.0	89.0	96.4	105.6	122.5	135.9

Source: Adapted from US DOE [52].

with 45% capacity [52]. In 2021, the top three countries with cumulative offshore wind installed were China, the United Kingdom, and Germany, with around 21.5, 12.2, and 7.8 GW, respectively. Also, both China and the United Kingdom have offshore wind projects under construction with over 6.8 GW capacity [55]. China and the United States are ranked first and second , respectively, in cumulative wind energy capacity globally. Also, Denmark, Ireland, Portugal, Spain, Germany, and the United Kingdom achieved a high level of wind energy penetration. Globally, cumulative offshore wind energy deployment could reach 177 GW by 2027, according to announcements by developers. As a result, many European countries have increased their deployment targets for offshore wind to reduce dependence on fossil fuels and address energy security [55].

The most common type of wind hybrid project combines wind power with storage technology, and other combinations are wind and PV, wind and gas, or wind, PV, and storage. In 2021, 41 wind hybrid plants were in operation in the United States, representing 2.4 GW of wind power and 0.9 GW of co-located resources [52]. Global offshore wind energy project data show that the farthest distance from shore is around 110 km, and the highest water depth is approximately 55 m by the end of 2021. The announced projects distance from shore and water depth will reach approximately 150 km and 75 m by 2027, respectively [55]. Offshore wind power generation is commercially competitive and can become cheaper than conventional power generation [56]. Airborne wind energy systems can be used in locations, such as islands or mines, that are not suitable for wind turbines [57].

1.2.1 Onshore (TRL 10)

Onshore wind power generation harnesses the kinetic energy of wind that is located on land to turn the blades of a wind turbine and run the power generator. Wind turbines consist of three blades mounted on a tall tower that can be rotated and adjusted their position to maximize their energy output (Figure 1.16). Onshore wind power generation is another renewable and sustainable method to provide a consistent and cost-effective power supply. Wind turbines can operate under different wind conditions and can be integrated with power grids to support electricity demand. This method is becoming more widespread and is used in many different settings, from small-scale residential wind turbines to large-scale wind farms.

Onshore wind power generation can be compatible with other forms of energy, such as hydrogen generation or fossil fuels. The performance of onshore wind power generation depends on wind speed, turbine design, and maintenance. These turbines have a high capacity and generate energy at a relatively low cost without GHG emissions or air pollutants during operation. This method can reduce dependence on fossil fuels and contribute to a more sustainable energy future. The

Figure 1.16 Example of onshore wind technology for power generation. Source: John Wiley & Sons.

main techno-economic factors are the location, size, and design of wind turbines, as well as the cost of materials, installation, and maintenance. Power generation via onshore wind turbines has been used for several decades and has a high TRL, around 10. The main benefits of onshore wind power generation are: (i) renewable and clean energy source, (ii) energy security and independence, (iii) minimal land use requirements, and (iv) cost-effective and low operational costs. The main drawbacks are: (i) visual and noise impacts, (ii) environmental impacts on wildlife and natural landscapes, (iii) wind speed variability, (iv) high capital costs, and (v) compatibility and integration challenges with existing energy systems.

Small wind farms can generate power for over 9,000 homes, and larger farms can capture wind energy more efficiently. The higher the turbines go, the windier and the more energy they can capture. Micro wind turbines can generate up to 20 TWh in different locations without centralized grid access by 2050. Onshore wind turbines can be used in rural areas to pump water, cook, or charge batteries, and reduce over 0.1 gigatons of GHGs. A small hybrid power system, including solar PV and onshore wind power technologies, has several benefits over a single system; for instance, we have more sun and less wind during the summer and vice versa during the winter. Therefore, hybrid systems are more beneficial for producing power when we need it at different times. It can be combined with engine generators, using diesel to produce power when neither solar nor wind systems provide energy. Also, it requires large enough battery storage to supply power for 1–3 days during the non-charging periods. Overall, onshore wind power generation is a sustainable, cost-effective, and reliable energy source. Still, it has some challenges related to visual and environmental impacts, intermittency, capital cost, and compatibility with existing energy systems. Further details have been provided by the U.S. DOE (2022) [52] and McKenna et al. (2022) [58].

1.2.2 Offshore (Two Technologies)

Offshore wind has the potential to generate a high amount (around 13,500 TWh/year) of renewable energy within 50 miles of coastlines, where over 50% of the U.S. population lives. Offshore wind turbines can be extremely tall to capture more energy because they can be transported by ships without any transportation limitations that land-based turbines have. However, working in the ocean has its own barriers, such as harsh weather. Currently, the first commercial offshore wind projects in the United States are off the coast of Rhode Island with 5 turbines total and 30 MW capacity that can power around 17,000 homes in New England, and the second one is in Virginia with 12 MW capacity. Generating 86 GW of offshore wind power by 2050 in the United States can create 160,000 jobs, cut GHG emissions by 1.8%, and reduce water consumption for power generation by 5%. In 2021, General Electric and Siemens Gamesa Renewable Energy had turbines with up to 14 MW capacity. Siemens and Vestas are the top offshore wind turbine manufacturers with over 25.1 and 8 GW capacity in 2021, respectively (Table 1.6). Globally, the energy cost for the U.S. fixed-bottom offshore wind projects is between $61 and $116 per MWh.

The top three global operating offshore wind structures are monopile, jacket, and high-rise pile caps with over 33.2, 5.8, and 3.2 GW capacity by the end of 2021,

Table 1.6 Annual U.S. market share of wind turbine manufacturers by MW between 2017 and 2021.

Turbine original equipment manufacturer	2017	2018	2019	2020	2021
GE wind	2,066	3,011	4,142	9,070	6,362
Vestas	2,481	2,886	3,003	6,023	3,499
SGRE	1,625	630	1,424	1,467	1,788
Nordex Acciona	806	866	549	451	1,765
Goldwind	6	171	14	202	0
Other	32	24	1	0	0

Source: Adapted from US DOE [52].

Figure 1.17 Schematic of offshore wind turbines for power generation. Source: National Renewable Energy Laboratory.

respectively (Figure 1.17). Also, the main announced offshore wind structures are monopile, semisubmersible, and jacket, with over 49.3, 14.1, and 11.8 GW capacity, respectively. Floating offshore wind turbines can be used in deep waters (above 60 m). The existing floating platforms are semisubmersible (mainly used, around 80% of the market share), spar buoy, tension leg, and barge. The total global pipeline for floating offshore wind energy grew to over 60 GW, and 123 MW of floating offshore wind projects were operational by the end of 2021 [55]. The forecasted offshore wind target in the United States will be between 26 and 32 GW by 2030.

Currently, the largest floating wind farm is Kincardine in Aberdeen, Scotland, with 47.5 MW capacity, using semisubmersible substructures. The global growth for floating offshore wind energy grew by around 125% in 2021, mainly attributed to South Korea, the United Kingdom, Brazil, and Australia. The latest reports show that the United Kingdom, South Korea, Spain, and Saudi Arabia are the top four countries with near-term floating offshore wind projects [55]. Floating offshore wind energy cost was around $200/MWh in 2021, and is estimated to reduce to $55/MWh by 2030 through optimizing the structure, sea construction, and supply chains, as well as accessing higher wind speeds. In addition, hydrogen, as the most abundant element on the earth, can be produced using offshore wind energy that allows transferring hydrogen to shore instead of power (electron).

Electrolysis can produce hydrogen by splitting water molecules into oxygen and hydrogen, using renewable energy sources, such as wind, with no or low carbon emissions. Currently, there is a high interest in offshore wind energy for producing hydrogen. However, a fraction of today's energy produces hydrogen from wind energy that can play an essential role in the global zero-carbon future.

1.2.2.1 Seabed-Fixed Offshore (TRL 9)

The science and principles of seabed-fixed offshore wind power generation are very similar to those of onshore wind power generation, harnessing the kinetic energy of wind to turn the blades of a wind turbine and its generator. Seabed-fixed offshore wind turbines are fixed to the ocean floor in shallow or deep water using a foundation (Figure 1.18). Power generation via offshore turbines is another sustainable and renewable method for providing a consistent and cost-effective power supply. Offshore turbines can be designed to operate under different wind conditions and can be integrated into the power grid to support power demand.

Offshore wind energy is becoming more widespread due to its benefits and applications from small-scale farms to large-scale parks. These turbines can provide power to homes, businesses, and communities, as well as support larger energy systems. Offshore wind power generation via seabed-fixed turbines is a secure and reliable energy source and is compatible with other forms of energy, including

Figure 1.18 Example of offshore wind technology for power generation. Source: John Wiley & Sons.

fossil fuels and renewable sources. Their performance depends on wind speed, turbine design, and maintenance. Offshore wind turbines have a higher capacity due to the higher wind speeds over water.

Generating energy from this technology is environmentally friendly due to no GHG emissions during the operation. The cost of seabed-fixed offshore wind power generation is typically higher than onshore wind power due to the challenges of installation, maintenance, and the more complex infrastructure needed offshore. This technology has a high TRL, around 9, which is still lower than onshore wind power. The main benefits are: (i) higher capacity and energy production than onshore wind power, (ii) no land use requirements, and (iii) low operational costs. The key drawbacks are: (i) high capital costs, (ii) complexity, (iii) potential environmental impacts on marine wildlife, (iv) potential impacts (e.g., visual and noise) for coastal communities, and (v) integration challenges with existing energy systems. Further details have been provided by the U.S. DOE (2022) [55].

1.2.2.2 Floating Offshore (TRL 8)

The basic science and principles of floating offshore wind power generation are very similar to fixed offshore wind power. The key difference is that floating offshore wind turbines are not anchored to the seabed, but are instead attached to floating platforms. Floating turbines can be located farther from shore in deeper waters, generating more power. They can provide a consistent and reliable power supply to remote areas and can also be integrated into existing power grids and energy systems to provide power to homes, businesses, and communities.

Similar to other offshore wind power generation systems, the performance of floating offshore wind power generation is influenced by wind speeds, turbine design, and maintenance. Earlier studies show that floating offshore wind turbines can generate power at a lower cost than fixed offshore wind turbines due to the availability of stronger and more consistent winds in deeper waters. However, the cost is higher compared to fixed offshore wind power due to the more complex infrastructure required for floating turbines. This method is environmentally friendly and does not release carbon emissions.

The TRL of this method is around 8, and it is expected to become more mature in the coming years. The main benefits include: (i) higher capacity than fixed offshore wind, (ii) no land use requirements, and (iii) potential for larger wind farms. The main drawbacks include: (i) high capital cost, (ii) complex installation and maintenance requirements, (iii) potential impacts on marine wildlife and coastal communities, and (iv) integration challenges with other energy systems. Overall, floating offshore wind power generation offers great potential as a sustainable and reliable energy source, but it still faces many challenges, including higher

capital costs and installation and maintenance requirements, potential environmental impacts, and integration challenges. More detailed information can be
found in the 2022 offshore wind market report by DOE [55].

1.2.2.3 Airborne (TRL 4)

Airborne wind system is a new technology for power generation through harvesting wind at higher altitudes, 400–600 m above the ground, using kites or tethered
flying devices (Figure 1.19). Similar to other wind power generation systems, this
method uses the kinetic energy of wind to generate mechanical energy and then
converts it into power. This method involves flying objects that are tethered to the
ground, with the energy generated from the movement of these objects. Airborne
wind power generation can be used in areas where traditional wind turbines are
not feasible.

Similar to other methods, the performance of airborne wind technologies
depends on wind speeds, flying objects' design, and maintenance. Airborne
wind power generation is a sustainable method that does not produce GHG
emissions and can help reduce dependence on fossil fuels and contribute to a
more sustainable energy future. The cost of airborne wind power generation is
currently higher than that of traditional wind power due to the relatively new and
complex infrastructure required for airborne turbines. The TRL for airborne wind
power generation is around 4, which still needs further investigation and testing
to become more mature in the coming years.

Airborne wind energy systems are still under development, with only a few
in operation. The benefits are fewer material requirements, easier transport and

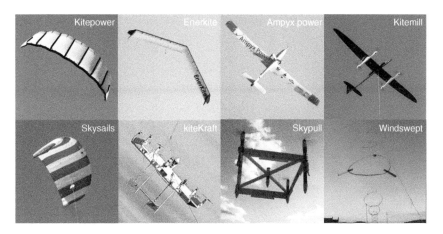

Figure 1.19 Examples of airborne wind technologies for power generation.
Source: Weber et al. [59]/National Renewable Energy Laboratory.

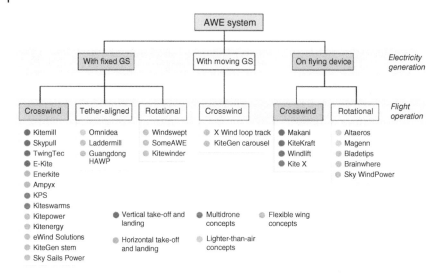

Figure 1.20 Airborne wind energy system classification. Source: Weber et al. [59].

installation, and greater operational flexibility that can generate higher energy yield with a lower carbon footprint compared to conventional wind turbines. Particularly, they can capture more constant, stronger winds at higher altitudes and adjust the altitude to more available winds. The main technical specifications are the type (soft-, fixed-, or hybrid-wing), the power generation (ground-gen or fly-gen), the wingspan and surface area, the altitude, and the power rate (Figure 1.20). The barriers are continuous automated operation (especially under extreme weather conditions) and social acceptance due to (i) the noise (sound emissions) emitted by the generators, (ii) the large size and visibility that has ecological effects on animals, such as birds if they use it in rural, remote areas, and (iii) the safety of this technology that can cause other damages if it crashes for some reasons [57]. More detailed information has been provided by the U.S. DOE (2021) [60].

1.3 Hydropower

Hydropower is another renewable energy source that generates affordable power from the domestic natural flow of water through impoundment (storing water in a reservoir), such as dams, diversion (channeling water of a river through a canal or pipe), or pumped storage hydropower (pumping water up to reservoir during low energy use). This method relies on the natural water cycle, driven by the

Figure 1.21 Example of power generation through hydropower technology. Source: John Wiley & Sons.

sun, to create the necessary force to generate power. Hydropower is produced by constructing a dam on a river or other body of water, creating a reservoir that stores water (Figure 1.21). When the water is released from the reservoir, it flows through a turbine and drives a generator to produce power.

Hydropower can generate power in both large- and small-scale applications in many countries worldwide as a primary or backup power source during periods of peak demand. Hydropower is compatible with other energy sources, and its performance depends on the amount of water available, the dam's height, and the turbine's efficiency. Like other renewable power generation methods, hydropower does not generate GHG emissions during its operation. The cost of hydropower can vary depending on the size and complexity of the project.

Hydropower technology is one of the oldest sources of power generation from a higher to lower elevation, using turbines and generators to convert this motion to power [61]. Hydropower is a well-established technology around the globe with TRL 11 that can seamlessly integrate with other energy sources. Currently, hydropower accounts for around 33–38% of renewable power generation and around 6.1–6.6% of total power generation (274 TWh) in the United States, using less than 3% of dams (out of 80,000 dams) for power generation, whereas other non-powered dams can generate around 4.8 GW power by 2050 [62]. Pumped storage hydropower is the largest U.S. energy storage technology, representing approximately 93% of all commercial long-duration storage capacity [63]. The latest studies focus on (i) increasing the efficiency of the hydropower facilities (e.g., turbines and generators), (ii) adding generators to non-powered dams that are used for irrigation or preventing floods, and (iii) exploring fish and natural habitat-friendly turbines and ladders that can let them swim around them [64, 65]. Small hydropower setups can generate over 1,000 TWh of energy by 2050 and cut over 3 gigatons of GHGs. In addition, hydropower requires careful design and installation to reduce negative ecological impacts, such as floods, fish migration, water quality, and sediment patterns.

The benefits of power generation through hydropower are: (i) renewable and clean energy source; (ii) high efficiency and reliability for continuous power generation; (iii) scalability; and (iv) other applications, such as irrigation or flood control. The main drawbacks are: (i) environmental impacts, including habitat destruction, water quality, aquatic ecosystems, and alteration of natural water flows; (ii) impacts on local communities and social structures; (iii) water dependency that can be affected by droughts and other environmental factors; and (iv) high capital and maintenance costs, especially in remote or difficult-to-access areas. Table 1.7 presents the average water consumption for different energy generation solutions [66]. More detailed information has been provided by the U.S. DOE (2021) [67] and Oak Ridge National Lab (2022) [65].

Table 1.7 Average water consumption for energy generation.

Solutions	Water consumption (L/MWh)
Biomass	85,100
Coal	2,220
Crude oil	3,220
Concentrating solar power	1,250
Geothermal	1,022
Hydropower	4,961
Natural gas	598
Nuclear	2,290
Solar PV	330
Wind	43

1.4   Geothermal

Geothermal power is a renewable energy source that can generate power continuously during the week in different weather conditions from hot air and water sources deep beneath the earth's surface. Geothermal power generation is based on the principles of thermodynamics and heat transfer. The heat from the earth is transferred to hot water or air to drive a turbine and generate power (Figure 1.22). The process involves drilling wells to access geothermal reservoirs and transferring hot water or steam to the power plants to drive a turbine for power generation.

This method is suitable for areas with high heat flow and accessible geothermal reservoirs, particularly areas with tectonic activity and high geothermal gradients. Geothermal power can provide baseload power and operate constantly to meet power demands. Geothermal power is a secure and reliable energy source that can achieve high efficiency and availability, and does not require external energy supplies. The performance of geothermal power plants is highly dependent on geothermal reservoir characteristics. Geothermal power generation is a sustainable method as it does not emit GHGs and has low environmental impacts compared to other forms of power generation. The construction and operation of geothermal power plants can still have environmental impacts on local ecosystems, and need to be managed properly. Geothermal power generation is a mature technology with high TRL. However, this method requires more innovation and

Figure 1.22   Example of geothermal power generation. Source: John Wiley & Sons.

improvements, such as reservoir management, drilling technologies, and power plant design.

Geothermal power generation is a relatively low-emission energy source compared to fossil fuels, but it can still have some environmental impacts. The construction of geothermal plants can disrupt local ecosystems and harm wildlife, and drilling wells and injection of fluids can potentially lead to induced seismicity. However, the impacts can be minimized if the plants are properly sited and managed. Geothermal power generation requires high capital costs due to the need for drilling and plant construction, but it can be cost-competitive over the long term due to low operating costs and stable fuel costs. Particularly, operational costs can be low, since the fuel source is free and the plants can operate 24/7 without the need for fuel storage or transport. The levelized cost of power for geothermal power is generally competitive with other renewable energy sources, such as wind and solar power. Geothermal power generation is a renewable, reliable, and dispatchable energy source that can provide baseload power. Geothermal power can also support local economic development by creating jobs and revenue in areas with geothermal resources. The construction and operation of geothermal power plants can have negative impacts on local ecosystems and communities, particularly if not managed properly. The sites for geothermal power plants are limited to areas with high heat flow, and the technology is not as widely available as other renewable energy sources. Geothermal can generate up to 3% of global energy by 2050 and cut emissions up to 10 gigatons of GHGs [68]. Among energy sources, geothermal has a smaller footprint, using less land per GWh, around $404\,m^2$, and coal, solar, and wind use 3,642, 3,237, and $1,335\,m^2$, respectively [69]. Geothermal power generation emits very small GHG emissions (around $50\,g\ CO_2$ eq. per kWhe), which is lower than solar power (four times less) and natural gas (6–20 times less).

Geothermal conditions do not exist on over 90% of the earth; however, we can create a similar condition in deep underground holes by adding water. Recent studies reported that over 39 countries could supply 100% of power demands from geothermal resources for both power supply and heating and cooling purposes. The environmental benefits are substantial, including low environmental impacts and emissions, and small physical footprints. According to the USGS, available geothermal resources in the United States (mainly in western states) can supply around 10% of the nation's energy needs and cut emissions from fossil fuels. Currently, geothermal energy provides approximately 60% of the power supply on the northern California coast. Also, geothermal energy recycles wastewater to generate power in California (Santa Rosa). The byproducts from solid wastes of the geothermal energy process (e.g., lithium) can be used in different industries and reduce the total cost of power generation. California and Nevada produce the most geothermal power, with over 2,600 and 790 MW production capacity,

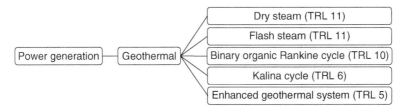

Figure 1.23 Main geothermal technologies for power generation.

respectively. Also, Utah, Oregon, Idaho, New Mexico, and Alaska can supply around 90, 38, 18, 15, and 1 MW of geothermal power, respectively [70]. Dry and flash steam formed the U.S. geothermal power generation; however, binary plants have been mainly used since 2000 due to the flexibility and use of lower temperature resources [70]. Figure 1.23 shows the main technologies for power generation, using geothermal resources.

The LCOE from geothermal processes is expected to reduce between $42 and $77 by 2050. The current techno-economic barriers to geothermal power generation are high exploration costs and risks for drilling, as well as the difficulty of creating reservoirs capable of sustained circulation for the enhanced geothermal system. Geothermal district heating systems use geothermal energy to warm up buildings near high-grade hydrothermal resources, which require high capital, particularly in the drilling and installation phases. Recent studies reported that China, Turkey, and Iceland have the highest geothermal district heating capacity worldwide, with 1,000, 53, and 52 MW thermal energy (MWth), respectively [70]. Overall, geothermal power generation has several benefits, including low emissions, reliable and dispatchable power, and local economic benefits. However, it has drawbacks, including high capital costs, limited availability, and potential negative impacts on local ecosystems and communities. More detailed information has been provided by the U.S. DOE (2022) [69] and NREL (2019) [70].

1.4.1 Dry Steam (TRL 11)

Dry steam geothermal technology is the most common process used for power generation from underground high-pressure (supercritical) steam with a temperature above 300 °C that flows directly to a turbine to run a generator and produce power (Figure 1.24). This technology can generate the highest amount of energy per fluid mass, directly from geothermal reservoirs without any separation systems. The science is based on the basic principles of thermodynamics and heat transfer. The heat from the earth is transferred to water in the geothermal reservoir, which turns into steam. This method requires access to geothermal reservoirs that produce dry steam, which can be used directly to drive a turbine.

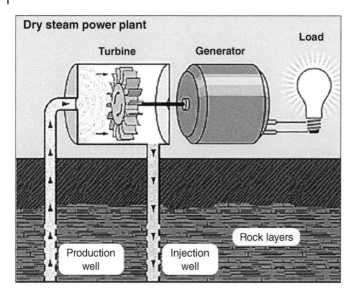

Figure 1.24 Schematic of dry steam technology for power generation. Source: U.S. Department of Energy, Energy Efficiency & Renewable Energy (public domain).

The steam is typically extracted from the reservoir using wells and then piped to the power plant. It is most suitable in areas where the geothermal reservoir produces dry steam and the resource is accessible through drilling wells. This method is compatible with the electricity grid and can operate 24/7 and meet a consistent demand for electricity. Dry steam geothermal power generation is generally considered a secure energy source as it is not dependent on external fuel supplies. Dry steam geothermal plants can achieve high levels of efficiency; however, their performance is highly dependent on the characteristics of the geothermal reservoir, and plant operators need to carefully manage the reservoir to ensure long-term sustainability. Dry steam geothermal power generation is a sustainable method with high TRL (around 11).

Dry steam geothermal power generation has relatively low environmental impacts compared to fossil fuel-based power generation. However, the construction and operation of geothermal power plants can still impact local ecosystems and communities, particularly if not managed properly. For example, drilling wells can lead to disturbance of the subsurface geology, and injection of fluids can cause induced seismicity. Additionally, some geothermal reservoirs contain naturally occurring gases that can be released into the atmosphere during power production. Earlier studies reported that direct use of geothermal fluids can release emissions and pollution (e.g., hydrogen sulfide, mercury, arsenic, and CO_2) that can contribute to potential eutrophication and acidification [71].

The latest life cycle assessment results show that geothermal power generation has less environmental impacts compared to fossil-based power generation, but it has more impacts compared to solar, wind, and hydropower plants. Particularly, acidification potential (SO_2 eq.) is similar (comparable) to fossil-based power generation [71]. Dry steam geothermal power generation has relatively high capital costs due to the need for specialized equipment and drilling operations. However, operational costs can be low since the fuel source is free, and the plants can operate 24/7 without the need for fuel storage or transport. The levelized cost of electricity for dry steam geothermal power is generally competitive with other renewable energy sources.

1.4.2 Flash Steam (TRL 11)

Flash steam geothermal technology pushes hot fluid from underground to a tank on the surface using a pump, which turns fluid into vapor quickly (flask vaporize) and drives a turbine and runs a generator to produce power (Figure 1.25). The basic science is very similar to dry steam technology, using the thermodynamic principle that when water is heated and rapidly depressurized, it will flash into steam. In this process, hot water from a geothermal reservoir is brought to the surface and passed through a high-pressure separator, where the sudden drop in pressure causes the water to flash into steam. The steam is then used to drive

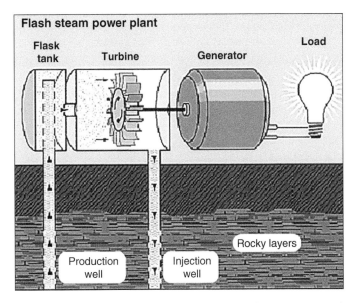

Figure 1.25 Schematic of flash steam technology for power generation. Source: U.S. Department of Energy, Energy Efficiency & Renewable Energy (public domain).

a turbine and generate power. There are three different flash systems: single, double, and triple, that have modified configurations, higher yield, and costs, respectively. The higher yield can justify the higher costs in several systems. Also, adding extra generators can increase efficiency. Flash steam geothermal power generation is a reliable and dispatchable source of renewable energy that can supply baseload power. The process of generating steam from hot water is relatively simple and efficient, and the technology has been used for several decades.

The benefits of flash steam geothermal power generation include: (i) reliability and dispatchability, (ii) low operating costs, (iii) the ability to provide baseload power, and (iv) job creation and revenue generation for communities. The main drawbacks are: (i) the need for suitable geothermal reservoirs that can be limited to specific geographic areas, (ii) the construction and operation of the power plant and its associated infrastructure can also have some environmental impacts, and (iii) challenges associated with long-term sustainability and preventing depletion of geothermal reservoirs.

Flash steam geothermal power plants require access to a geothermal reservoir containing hot water at a temperature of around $180\,°C$ ($356\,°F$). The sites are limited to areas with high heat flow, making finding suitable locations difficult. The plants can be integrated with existing power grids and other renewable energy sources (e.g., wind and solar power) to provide a stable and reliable power source. These plants have relatively low security risks compared to other forms of power generation since the fuel source is located underground, and the technology does not produce hazardous waste. Flash steam power plants can have high capacity factors and availability since the fuel source is stable and predictable. The performance depends on the temperature and flow rate of the geothermal reservoir. Power generation via flash steam geothermal is a sustainable energy source with low environmental impacts and emissions, and it can be used indefinitely as long as the geothermal reservoir is managed properly. Flash steam geothermal has been used commercially for several decades and is a mature (TRL 11) and highly efficient and reliable solution for power generation.

The environmental impacts of this method include the release of some gases and minerals from the geothermal reservoir, such as hydrogen sulfide, ammonia, boron, and arsenic. These gases and minerals can harm local ecosystems and communities. Also, the construction of power plants and their associated infrastructure can impact the natural environment. Techno-economic studies show the cost of a flash steam geothermal power plant depends on the plant size, required equipment, geothermal reservoir characteristics, and materials. The operating costs are relatively low due to the free fuel source and low maintenance costs. The initial capital costs are high due to implementation challenges, as well as the need for specialized equipment and infrastructure to extract and utilize the steam. The levelized cost of power from flash steam geothermal depends on the location and other factors, but it is competitive with other renewable energy sources.

1.4.3 Binary Organic Rankine Cycle (TRL 10)

Binary organic Rankine cycle (ORC) geothermal power generation is based on the principles of thermodynamics, which is the relationship between heat, work, and energy. The system uses a heat exchanger to transfer heat from a geothermal fluid to a working fluid, which is typically an organic compound. The working fluid is then vaporized, driving a turbine to generate power. Binary cycle technology uses a heat exchanger and two fluid types: hot fluid from underground and heat transfer fluid (Figure 1.26). The heat transfer fluid has a much lower boiling point than the hot fluid that can quickly turn or flash to vapor at a lower temperature and spin a turbine that powers a generator and produces power. ORC can generate power from low- and medium-temperature heat, e.g., geothermal sources. ORC technology uses organic fluids that are appropriate for medium temperatures between 100 and 220°C. Some of ORC fluids are isopentane,

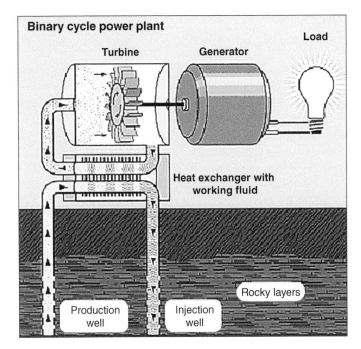

Figure 1.26 Schematic of binary cycle technology for power generation. Source: U.S. Department of Energy, Energy Efficiency & Renewable Energy (public domain).

isobutene, *n*-pentane, and mixed isopentane and isobutane. ORC systems have been utilized for emission-free power generation from geothermal sources for many years.

Binary ORC for geothermal power generation typically includes a heat exchanger, a vaporizer, a turbine, a generator, and a condenser. It is suitable for a wide range of geothermal reservoirs, including those with lower temperatures and flow rates than those required for other types of geothermal power plants. The system can also be used in areas with limited water resources since it does not require large amounts of water for cooling. Binary ORC can be integrated into the existing power grid, making it compatible with other energy sources. The system can also be designed to operate in tandem with other renewable energy sources. Their performance mainly depends on geothermal fluids' characteristics, temperature, and flow rate, as well as the efficiency of heat exchangers, turbines, and generators. Binary ORCs for geothermal power generation can provide high efficiency and operate at different geothermal fluid temperatures. Binary ORC is a sustainable, mature technology with TRL 10 that has been successfully implemented in several geothermal power plants worldwide.

Binary ORCs have relatively low environmental impacts compared to fossil fuel power generation. The technology does not emit GHGs, and the geothermal fluid used in the process is typically reinjected back into the reservoir. However, binary ORC systems, like all geothermal power plants, can impact local hydrology, geology, and ecology, requiring careful consideration of these factors during site selection and development. Binary ORC systems are generally considered a competitive option for power generation due to a relatively high capacity factor, which is the percentage of time the plant operates at full capacity, and low operating costs due to the stable and predictable nature of geothermal resources. However, the capital costs can be high, and the technology is not suitable for all geothermal reservoirs. Particularly, the capital costs for a binary ORC geothermal power plant are typically higher than those for conventional fossil fuel power plants of similar capacity. However, the operational costs are relatively low due to the stability and predictability of geothermal resources. The operating costs include maintenance and replacement of equipment, labor, and fuel. Latest studies demonstrated that geothermal-based ORC performance highly depends on the fluid and operating conditions [72, 73]. Binary ORC systems have several benefits, such as reliability, stability, and sustainability, as well as operating at a wide range of geothermal fluid temperatures that make them suitable for a range of geothermal reservoirs. The systems are also compact and can be built in relatively small sizes. Binary ORCs have some drawbacks, such as geothermal fluid requirements and deployment barriers (e.g., high capital cost), as well as regular maintenance and replacement of equipment that can add to operational costs.

1.4.4   Kalina Cycle (TRL 6)

Kalina cycle is another method for power generation from geothermal sources and is based on the principles of thermodynamics and heat transfer. The working fluid is evaporated in a low-pressure turbine, which expands the fluid and drives a generator to produce power. Kalina cycles extract heat from the geothermal fluid in a heat exchanger. The geothermal fluid heats the ammonia–water mixture, which vaporizes and expands through a turbine to generate power. The spent fluid is then condensed, and the process is repeated.

Kalina cycle technology is suitable for areas with medium- to high-temperature geothermal resources due to high flexibility with different geothermal fluid temperatures and flow rates. The benefits include the ability to: (i) integrate with existing geothermal power plants or as a stand-alone power generation system, (ii) provide a reliable and consistent power supply, (iii) operate continuously with minimal maintenance and downtime, (iii) have high efficiency due to the use of an ammonia–water mixture that has a lower boiling point than other fluids used in traditional binary systems, and (iv) generate power at lower temperatures, which are more common in geothermal resources. Organic Rankine and Kalina cycles are the standard binary cycle systems with 10–13% thermal efficiency. Kalina cycle is a sustainable technology with no GHG emissions during operation, and the geothermal resource is replenished naturally over time. This technology has TRL 6, has been successfully deployed in several geothermal power plants around the world, and has a proven track record of reliability and performance.

The environmental impacts of Kalina cycle systems are very similar to those of other types of geothermal power generation, which are significantly lower than those of conventional fossil fuel power plants. However, the drilling and construction operations can impact local hydrology, geology, and ecology. Kalina cycles can be a competitive option for power generation due to higher efficiencies than other geothermal power generation systems. Earlier techno-economic studies show the capital cost is high due to complexity and suitability for all geothermal reservoirs, and the operational costs are relatively low due to the stable and predictable nature of geothermal resources, which includes maintenance and replacement of equipment, labor, and fuel. Further details about organic Rankine and Kalina cycles have been provided by [74].

1.4.5   Enhanced Geothermal System (TRL 5)

Enhanced geothermal system (EGS) uses engineered reservoirs to extract heat and steam by injecting water to underground to be in contact with hot rocks and transferring heat to the surface for power generation (Figure 1.27). EGS uses heat extracted from rocks, reservoirs or other geothermal sources (e.g., natural

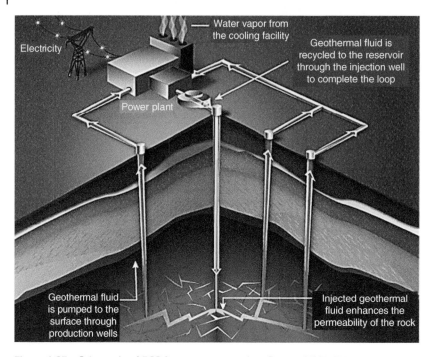

Figure 1.27 Schematic of EGS for power generation. Source: NREL [75].

hydrothermal systems) that are not naturally porous and permeable by creating artificial fractures and circulating water through these fractures to extract the heat. The basic science behind EGS is based on the principles of geology, drilling, and fluid dynamics. EGS technology works by drilling deep wells into the earth's crust and injecting water under high pressure into the rock formation to create artificial fractures.

The benefits of EGS technology include the ability to: (i) work in areas with low-to high-temperature geothermal resources, (ii) integrate with existing geothermal power generation systems or as a stand-alone technology, (iii) combine with other renewable energy sources to provide a reliable and consistent power supply, and (iv) operate continuously with minimal maintenance and downtime. EGS is secure, reliable, and sustainable to generate a significant amount of power, but its efficiency is currently lower than traditional geothermal power generation systems. The geothermal resource is replenished naturally over time, but the water used in the process may contain impurities that need to be treated before disposal. Other drawbacks include that it requires special drilling techniques and is only economically feasible in areas with suitable geological conditions.

Earlier environmental impact assessment studies show that EGS has lower environmental impacts compared to fossil fuel-based power generation, as they do not release GHG emissions during operation. The main environmental concern is due to significant water requirements for the injection and extraction processes that can impact local water resources. Also, the creation of artificial fractures in the rock formation can cause seismic activity in some cases. The techno-economic analysis shows that EGS requires high capital costs, primarily due to the cost of drilling and completing the wells, and the associated equipment required to create and maintain the artificial reservoirs. Once the system is operational, the fuel costs are low, making EGS economically competitive with other renewable energy sources. The operational costs are typically low, as the fuel source is renewable and does not require continuous fuel purchases. Also, EGS has a long lifespan, around 30 years or more.

EGS technology has TRL around 5, which is still under development and has not yet been widely deployed commercially. Currently, the main focus is to reduce capital costs by improving site selection and drilling techniques. Overall, EGS is a promising technology to generate a significant amount of power in a wide range of geothermal conditions, but it requires lower capital costs and higher efficiency to become commercially viable. More detailed information about EGS has been provided by NREL [75].

Dry and flash steam geothermal power generation are well-established technologies with high efficiency and low capital costs. But they are limited to specific geologic conditions and have high environmental impacts. Binary organic Rankine and Kalina cycles are more advanced technologies that can generate power at lower temperatures and with lower environmental impacts, but have higher capital costs and require more advanced equipment. EGS is the most technically complex of these technologies and requires the most advanced drilling and reservoir engineering, but has the potential to generate power in a wide range of geological conditions. In terms of TRL, dry steam and flash steam geothermal power generation are the most mature technologies, with well-established projects around the world (Table 1.8). Binary organic Rankine and Kalina cycles are still relatively new technologies, with fewer commercial-scale projects, while EGS is still in the research and development stage, with a limited number of pilot projects.

1.5 Hydrogen

Power generation from hydrogen involves the use of hydrogen as a fuel to generate power and typically involves two main steps: (i) the production of hydrogen and (ii) the conversion of hydrogen into power. The primary methods of hydrogen production include steam methane reforming, electrolysis, and gasification. Steam

Table 1.8 Comparison of different geothermal technologies for power generation.

Technology	Mechanism
Dry steam	This technology uses natural steam from underground reservoirs to drive a turbine and generate power. The steam is produced when water is heated by the geothermal heat source and turns into steam. The steam is then directly fed into a turbine to generate power
Flash steam	This technology involves drilling wells into geothermal reservoirs and bringing hot water and steam to the surface. The hot water and steam are then separated in a flash tank, with the steam driving a turbine to generate power
Binary ORC	This technology involves pumping hot water from the geothermal reservoir through a heat exchanger, where it heats a secondary fluid with a lower boiling point. The secondary fluid vaporizes and drives a turbine to generate power
Kalina cycle	This technology is similar to binary organic Rankine cycle, but uses a mixture of ammonia and water as the secondary fluid. The mixture has a variable boiling point, allowing for more efficient power generation at lower temperatures
EGS	This technology involves drilling deep wells into hot rock formations and creating an artificial reservoir. Cold water is pumped into the reservoir, and the heat from the surrounding rock heats the water to produce steam, which is then used to generate power

methane reforming is the most common method of hydrogen production, where natural gas is reacted with steam to produce hydrogen gas and carbon dioxide. Electrolysis involves splitting water into hydrogen and oxygen using an electric current (Figure 1.28). Gasification is another method that uses high temperatures to convert feedstocks (e.g., biomass or coal) into hydrogen. The technologies for converting hydrogen into power include fuel cells and combustion engines. Fuel cells are electrochemical devices that convert the chemical energy of hydrogen into power. Fuel cells have high efficiency and low emissions, making them ideal for power generation. Combustion engines are another option that involves burning hydrogen in an internal combustion engine to generate power.

Hydrogen could be a key option to reach net-zero emission targets across multiple energy sectors in the future, using renewable energy sources (e.g., solar, wind, biomass, hydropower, and geothermal). Hydrogen reacts and binds with other elements, which requires energy to separate it from other elements. In 2022, the United States produced over 10 million metric tons of hydrogen per year, around 14% of global supply [76]. Globally, the primary method and energy source for hydrogen production is using natural gas (over 95%)

Figure 1.28 Schematic of electrolysis for power generation from hydrogen. Source: John Wiley & Sons.

due to its low cost; however, it releases GHG emissions and requires carbon capturing and sequestration processes, which are not fully commercialized yet. H_2-rich fuel from carbon capture, utilization, and sequestration (CCUS) technologies can be used in a Rankine and Brayton combined cycle plant for power generation. Additionally, hydrogen can play a key role in the near future in storing excess renewable energy that would be otherwise wasted. Battery storage will not be cost-effective for addressing high seasonal storage for a longer duration. Converting the renewable energy to hydrogen and then converting back to power with hydrogen-driven turbines or fuel cells shows huge potential. Besides, hydrogen has several applications in producing transportation fuels, ammonia, methanol, and steel that can reduce the GHG emissions of these industries. Since hydrogen is a highly flammable gas, it requires special handling and storage. Some of the most common methods of hydrogen storage include compressed hydrogen gas, liquid hydrogen, and solid-state hydrogen storage. Currently, the major challenges for hydrogen are high flammability compared with other fuels and low conversion efficiency with electrolysis, as well as storage and transportation requirements. Particularly, bulk hydrogen transportation requires specific pressurizing, cooling, and investments. Overall, power generation from hydrogen offers a promising pathway to decarbonize the energy sector and reduce GHG emissions. However, the existing technologies are still in the early stages of development and face several technical and economic challenges that must be overcome before they can be widely adopted (Figure 1.29).

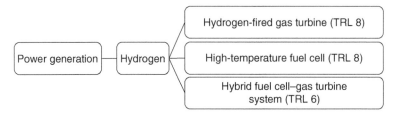

```
                          ┌──────────────────────────────────────┐
                          │ Hydrogen-fired gas turbine (TRL 8)    │
                          └──────────────────────────────────────┘
┌──────────────────┐   ┌──────────┐   ┌──────────────────────────────────────┐
│ Power generation ├───┤ Hydrogen ├───┤ High-temperature fuel cell (TRL 8)    │
└──────────────────┘   └──────────┘   └──────────────────────────────────────┘
                          ┌──────────────────────────────────────┐
                          │ Hybrid fuel cell–gas turbine          │
                          │         system (TRL 6)                │
                          └──────────────────────────────────────┘
```

Figure 1.29 Main hydrogen technologies for power generation.

1.5.1 Hydrogen-Fired Gas Turbine (TRL 8)

Hydrogen-fired gas turbine systems use hydrogen as the primary, highly flammable gas that can be burned to run the gas turbine and convert the hydrogen energy to mechanical energy for heat and power generation (Figure 1.30). The chemical reaction in burning hydrogen is highly exothermic, releasing a large amount of energy in heat. Particularly, this process has four steps: (i) air is drawn into the compressor, where it is compressed and heated; (ii) the compressed air is then mixed with hydrogen and burned in the combustor; (iii) the hot combustion gases expand through the turbine, where they drive the rotation of turbine blades, generating mechanical energy; and (iv) the exhaust gases then exit the turbine and are released into the atmosphere.

The efficiency and emissions of a hydrogen-fired gas turbine depend on the turbine design, combustion process, and fuel quality. Hydrogen gas has a high energy density and burns cleanly without carbon emissions. However, hydrogen combustion can release NO_x if the temperature is too high. Hydrogen-fired gas turbines may use low-NO_x combustors or exhaust gas recirculation systems to mitigate

Figure 1.30 Example of hydrogen gas turbine with 30 MW capacity. Source: Japan's Kawasaki Heavy Industries Kawasaki Heavy Industries [77].

NOx emissions. Gas turbines have been used widely for power generation due to their efficiency and flexibility in integrating with other systems, such as Rankine or Brayton cycles. Hydrogen turbines work very similar to internal combustion engines, using fossil fuels (e.g., coal or natural gas) for power generation; however, they are less efficient (around 20–25%) than gasoline combustion engines due to their low volumetric energy density [78]. The key factors are the applied fuel, efficiency, and emission. Recent studies investigated mixed fuels, such as natural gas and hydrogen, on hydrogen-fired gas turbines that can significantly affect the exergy efficiency and reduce carbon emissions due to improvement in mixing fuel and air [79]. Overall, a hydrogen-fired gas turbine can provide a high-efficiency and low-emission power generation option for the energy sector.

1.5.2 High-Temperature Fuel Cell (TRL 8)

Compared to hydrogen combustion engines or turbines, fuel cells can convert the chemical energy of hydrogen to power directly, and their conversion efficiency can reach to 80% [78]. Fuel cells are commercially applied and have been used in various industries, such as transportation. The operation principle of fuel cells is based on supplying hydrogen and oxygen that pass through the anode and cathode sides, respectively, where H^+ ions and electron are released, and water is produced from mixing proton and oxygen (Figure 1.31). Fuel cells have different types and components that can facilitate the process, such as temperature or catalyst. The common catalyst is carbon materials coated with platinum, but the overall reaction is as follows:

$$2H_2 + O_2 \rightarrow 2H_2O + \text{electricity} + \text{heat}$$

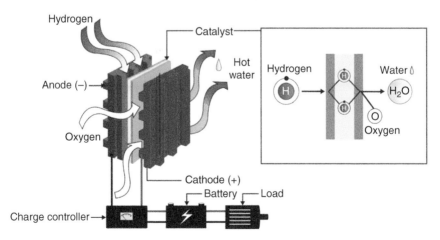

Figure 1.31 Power generation through hydrogen fuel cells. Source: Wiley Digital Library.

High-temperature fuel cells operate at temperatures between 600 and 1,000 °C, which is much higher than the operating temperatures of other fuel cells, such as proton-exchange membrane fuel cells and alkaline fuel cells. High-temperature fuel cells offer several benefits over other fuel cell technologies, such as efficiency, fuel flexibility, durability, and low emissions. Particularly, they can operate on a wide range of fuels, including hydrogen, natural gas, and biofuels, and achieve efficiencies up to 60%, making them more efficient than other types of fuel cells. The high operating temperatures make them more durable than other fuel cell technologies. Also, they release low emissions, making them an attractive power generation option. Recently, high-temperature fuel cells gained more attention due to several benefits compared to low-temperature fuel cells, such as efficient water and heat management, enhanced reaction kinetics, and simple plate design [80]. In terms of power generation from hydrogen, high-temperature fuel cells use a solid oxide electrolyte to separate the hydrogen ions and electrons. The hydrogen ions move through the electrolyte to the cathode, where they react with oxygen to produce water. The electrons, on the other hand, move through an external circuit to generate power. They have several applications, including power generation for homes, businesses, and large-scale power plants. They are also used in the transportation sector to power vehicles, such as buses and trucks. However, high-temperature fuel cells are still relatively expensive and require further development to address the degradation of catalyst and membrane at high temperatures, as well as maintaining the high operating temperature for the long-term process.

Additionally, phosphoric acid fuel cells can operate at around 180 °C with a stack voltage efficiency of 80%. The benefits are high efficiency with heat co-generation; however, the drawbacks are high catalyst cost and low current density. Proton-exchange membrane fuel cells can operate at 80–100 °C (low temperature) or 200 °C (high temperature) with a stack voltage efficiency of 50–60%. This fuel cell is widely used in different applications, but the expensive catalyst is the drawback. Solid oxide fuel cells operate at 800–1,000 °C with a stack voltage efficiency of 60–80%. The benefits of this fuel cell are reusable heat and lower cost, but the drawback is the issues with metal corrosion. Alkaline fuel cells operate at around 70 °C with a stack voltage efficiency of 60%. The benefits are good current response and its space applications. Molten carbonate fuel cells operate at around 650 °C with stack voltage efficiency of 60–80%. The main benefits are high conductivity and current density, which can be used in large-scale stationary applications. Fuel cells can be used for CHP generation or cold and power generation (co-generation), as well as cold, heat, and power generation (tri-generation). The main challenges are fuel cell efficiency and durability, which led to reduced competitiveness due to high operational costs.

1.5.3 Hybrid Fuel Cell–Gas Turbine System (TRL 6)

Hybrid fuel cell–gas turbine systems use hydrogen for power generation through fuel cells, and the waste heat from this process can be used to drive gas turbines and generate power. The main steps include: fuel cell operation and waste heat recovery, and the key principles are based on electrochemical and thermochemical processes. Particularly, hydrogen is fed into the anode side of a fuel cell, where it reacts with a catalyst and is split into protons and electrons. The protons move through an electrolyte membrane to the cathode side, while the electrons move through an external circuit to generate power. The waste heat generated during the fuel cell operation is used to drive a gas turbine, which expands and drives a generator to produce additional power.

Similar to other processes, the key parameters are efficiency and power output. Combining the fuel cell and gas turbine in the hybrid system and waste heat recovery can result in higher overall efficiency (over 10%) than using either technology alone. The power output depends on the system size, fuel cell type, and gas turbine type. Hybrid fuel cell–gas turbines can range from small-scale systems for residential or commercial use to large-scale systems for industrial applications. Earlier studies investigated hybrid fuel cell–gas turbine systems for combined power and heat generation, and their results show that these systems have high process efficiency and flexibility regarding the fuel type.

The commercialization of hybrid systems highly depends on the cost of materials. Currently, most hybrid systems use methane due to its lower cost and ease of handling compared to hydrogen. Overall, hybrid fuel cell–gas turbine systems offer a sustainable and reliable alternative to traditional power generation methods,

Table 1.9 Comparison of high TRL technologies for power generation from hydrogen.

Technology	Benefits	Drawbacks
Hydrogen-fired gas turbines	High efficiency (around 60–65%), low emissions (no carbon, only water vapor), and low maintenance	High cost (due to high hydrogen production cost), safety (hydrogen flammability), special handling (hydrogen storage and transportation)
High-temperature fuel cells	High efficiency (around 50–60%), low emissions, high scalability, and low maintenance	High cost and short lifespan (5–10 years)
Hybrid fuel cell–gas turbine systems	High efficiency (around 70–75%), low emissions, and high scalability	High cost and system complexity

with lower GHG emissions, higher energy efficiency, and reduced water usage. However, the high cost and complexity of the system currently limit its widespread adoption. More detailed information about various hybrid fuel cell-gas turbine systems has been provided by [81]. Additionally, each technology has its own benefits and drawbacks, and technology selection depends on various factors, such as hydrogen availability, project scale, and total cost (Table 1.9).

1.6 Marine

Marine-based energy is a renewable source from the natural movement of water (either horizontal or vertical) in the world's oceans or rivers. Particularly, marine energy technologies (e.g., buoys) capture the kinetic energy of waves, tides, currents, and thermal energy for power generation that can be fed into the grid or used directly by coastal communities or offshore facilities (Figure 1.32). These technologies can be deployed in various ocean environments, including shallow and deep waters. The usability of these technologies depends on the availability and intensity of the ocean resource, the distance from the shore, the grid connection, and the local energy demand. Marine technologies for power generation can be compatible with various types of energy systems, including traditional fossil fuel-based power plants and renewable energy systems. The compatibility of these technologies with other energy systems depends on the local energy infrastructure and the electricity demand. They are generally secure, as they are designed to withstand harsh ocean conditions, such as storms, waves, and currents. However, these technologies may face security challenges related to data privacy, cyberattacks, and piracy.

Marine energy is geographically diverse and abundant throughout the world and can generate over 60% of the power needs and cut over 1 gigaton of GHGs by 2050. The performance of marine technologies for power generation depends on various factors, such as design, materials used, ocean conditions,

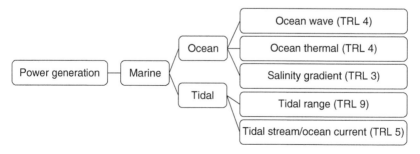

Figure 1.32 Major marine technologies for power generation.

maintenance, and grid connection. These technologies have different performance characteristics, such as capacity factor, efficiency, reliability, and lifetime. The TRL of marine technologies for power generation varies depending on the maturity of the technology, the number of pilot projects, and the commercialization stage. Some technologies, such as tidal energy and offshore wind, have a high TRL and are already being deployed commercially, while others, such as ocean thermal energy conversion, are still in the pilot stage. Their drawbacks are high installation, operation, and maintenance costs in harsh environments compared with other methods (e.g., solar and wind).

Marine technologies for power generation have varying environmental impacts, depending on technology, location, and the scale of deployment. These impacts can include noise pollution, visual impacts, habitat disturbance, and marine mammal and fish mortality. However, compared to traditional fossil fuel-based power plants, marine technologies generally have lower GHG emissions and other air pollutants. The capital and operational costs of these technologies vary depending on the scale, capacity, and location. Generally, these technologies have higher upfront costs than traditional fossil fuel-based power plants but lower operational costs over their lifetime. These technologies have several benefits, such as the ability to generate power from renewable sources, the potential to provide energy security and independence, the ability to create local jobs and economic development, and the potential to reduce GHGs. The main drawbacks are high capital costs, ocean resource limitations, integration with grid connection, energy storage, and potential conflicts with other ocean uses, such as fishing and recreation. Overall, marine energy resources are excellent options to complement other renewable resources due to their daily and seasonal availability. Also, marine energy technologies can be a game changer in turning seawater into clean drinking water.

1.6.1 Ocean (Three Technologies)

Ocean-based energy is a renewable and resilient solution for power generation from oceans and waves, which is based on the principle of converting the kinetic or thermal energy of the ocean into power. The science behind ocean power generation technologies involves various disciplines, such as fluid mechanics, materials science, electrical engineering, and environmental science. Wave energy conversion science involves understanding wave characteristics (e.g., amplitude, frequency, and direction) and wave energy devices for capturing and converting wave motion to mechanical energy that can be converted into power using generators or other methods.

Ocean thermal energy conversion science involves understanding the temperature differences between warm surface waters and cold deep waters, and thermal

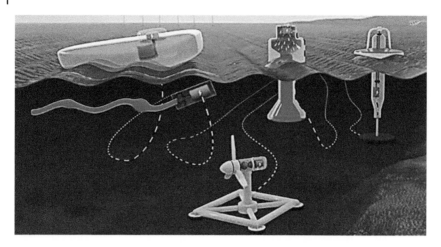

Figure 1.33 Examples of ocean technologies. Source: National Renewable Energy Laboratory [82].

energy devices that can extract and convert the thermal energy to mechanical energy for power generation using generators. Overall, the science of ocean power generation involves understanding the unique characteristics of each ocean energy source and developing devices that can efficiently and reliably capture and convert the energy to power.

Globally, the cities close to ocean coastlines and river resources can use this solution to generate power near where it is needed with short transmission lines. Ocean energy can provide power to remote and island communities. The technology readiness of ocean-based energy generation technologies is still at the prototype stage due to the harsh environment, social acceptability, and their effects on marine mammals, fish, and birds (Figure 1.33). Large-scale flow changes can disrupt the natural ecosystem, affect sediment transport, and cause water quality degradation.

1.6.1.1 Ocean Wave (TRL 4)

The interaction between ocean surface water and wind (due to temperature changes by the sun) creates waves. Ocean wave energy technologies can generate power from the movement of surface waves through different wave-capturing devices, such as (i) floating- or submerged-point absorbers, (ii) shore- or floating-based oscillating water columns, (iii) shore- or floating-based over-topping devices, (iv) attenuators, and (v) oscillating wave surge converters (Figure 1.34).

Point absorber wave energy converters are technologies that use buoy-like structures to capture the energy of ocean waves. The movement of the buoy drives a

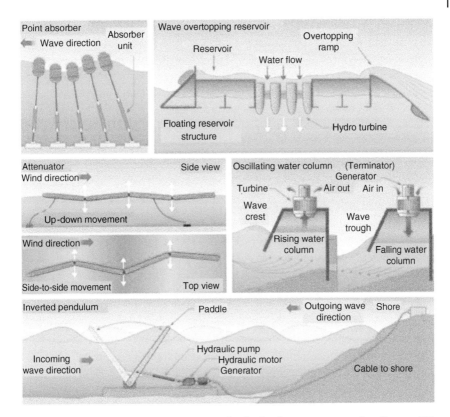

Figure 1.34 Schematic of wave energy technologies for power generation. Source: U.S. Department of Energy and National Renewable Energy Laboratory (public domain) [83].

generator, which produces power. Several point absorber wave energy converters have been deployed around the world, such as Pelamis Wave Energy Converter in Portugal and WaveRoller in Finland. Oscillating water column converters use the motion of waves to compress and decompress air in a chamber. The movement of the air drives a turbine, which generates power. An example of an oscillating water column wave energy converter is LIMPET device in Scotland. Overtopping converters use the potential energy of water stored in a reservoir to drive a turbine. As waves enter the reservoir, they fill the reservoir and increase the potential energy of the water. When the water is released, it flows through a turbine, generating power. An example of an overtopping wave energy converter is the Wave Dragon device in Denmark. Attenuator converters consist of multiple segments that are connected by hinges. The motion of the waves causes the segments to move, which drives hydraulic pumps that generate power. An example of an attenuator wave energy converter is the CETO device in Australia.

The main principle of wave energy is that when the wave passes and drives the fluid and hydraulic pump, motor, turbine, and generator, its can convert mechanical energy into power. Ocean wave converters utilize various mechanical, hydraulic, or pneumatic systems to generate power. They are usually deployed in offshore locations where there are consistent waves. They can be installed as individual devices or as part of a larger array. Converters must be compatible with the harsh and corrosive ocean environment to withstand unpredictable and powerful waves and forces. Wave energy converters can be used in remote offshore locations to provide a reliable renewable energy source. Their performance can be affected by wave height, direction, and frequency, as well as the design, maintenance, and operating conditions. Wave converters have reached TRL 4, with several devices already in the testing phase.

The environmental impacts of wave energy converters include habitat disruption, noise pollution, and collisions with marine mammals or other sea life. These impacts are considered less severe compared to the associated GHG emissions from fossil fuel-based power generation. Also, these converters have high capital costs due to the complexity and durability requirements and low operating costs since ocean wave is free. Like other technologies, their benefits are reliability and sustainability since they do not release GHG emissions, along with power generation using domestic resources. The drawbacks are high capital costs, deployment barriers, variability, and unpredictability of ocean waves that can affect the efficiency of this technology. These technologies have been tested and demonstrated at various scales, and several pilot projects have been deployed and have shown promising results in terms of efficiency, reliability, and environmental impacts. Further information about ocean wave energy has been provided by [84].

1.6.1.2 Ocean Thermal (TRL 4)

Ocean thermal energy can be harnessed from temperature differences (up to 25 °C) in sun-warmed surface water and cold deep ocean water through thermal energy conversion. The existing technologies use warm surface seawater, heat exchanger, and fluid with low boiling point (e.g., ammonia) to vaporize and run the turbine and generator, then cold deep seawater and condenser turn vapor into liquid to reuse throughout the conversion process. Ocean thermal power generation is based on the scientific principle of the thermodynamic cycle, which involves the transfer of heat energy from one body to another to produce mechanical work. The process of ocean thermal power generation involves the following basic steps:

- Collecting warm surface water from the ocean using a pipe or other collection device
- Vaporizing a working fluid (e.g., ammonia or a fluorocarbon) in a heat exchanger to absorb heat energy from the warm water and vaporize it into a gas

- Running a turbine to produce mechanical energy that can be converted into power
- Condensing the working fluid using cold water from the ocean depths to condense it back into a liquid that can be recycled back to the heat exchanger to repeat the cycle

This process is repeated continuously as long as there is a temperature difference between the warm surface water and the cold deep water. The greater the temperature difference, the more efficient the process becomes. The efficiency of the process is also affected by the design and efficiency of the heat exchanger and the turbine. There are several ocean thermal conversion technologies, such as closed-, open-, or hybrid-cycle systems (Figure 1.35). For example, the closed-cycle ocean thermal technology uses a closed-loop system that circulates a working fluid, such as ammonia or refrigerant, between a heat exchanger and a turbine. Warm surface water is used to vaporize the working fluid, which drives the turbine, producing electricity. Cold deep ocean water is then used to condense the vapor back into a liquid, completing the closed loop.

Ocean thermal power generation can be used in areas with a large temperature difference between surface water and deep water, such as tropical and subtropical regions. The system can produce power continuously as long as the temperature difference is maintained. Ocean thermal converters may not be compatible with certain marine ecosystems and fisheries. The performance of ocean thermal

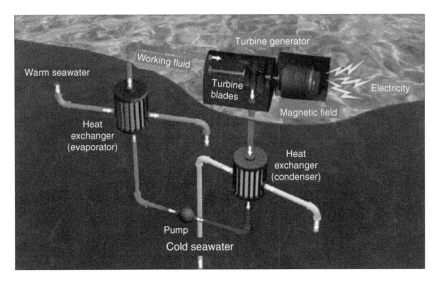

Figure 1.35 Schematic of a closed Rankine cycle ocean thermal energy conversion system. Source: Ascari et al. [85]/U.S. Department of Energy/Public Domain.

power generation system highly depends on the temperature difference between the surface water and deep water, as well as the efficiency of the heat exchanger and the heat engine. This system has a low conversion efficiency compared to other renewable energy sources. Ocean thermal power generation is still in the early stages of development, with several demonstration projects currently underway. Ocean thermal power generation can have environmental impacts, such as changes in water temperature and currents, noise pollution, and impacts on marine ecosystems and fisheries.

The benefits include its ability to provide a stable and reliable power supply, low emissions, and its potential to work with other renewable energy sources. The drawbacks include its high capital and operational costs, low conversion efficiency compared to other solutions, and potential environmental impacts on marine ecosystems and fisheries. The major challenge is accessing large volumes of warm and cold water with at least 18 °C thermal differences. More detailed information about ocean thermal energy has been provided by [86].

1.6.1.3 Salinity Gradient (TRL 3)

Salinity gradient technologies can generate power from the difference in salt concentration between freshwater and seawater. It is based on the principle of reverse electrodialysis, a process that involves the transfer of ions across a semi-permeable membrane to generate an electric potential difference, which involves the following basic steps:

- Collecting seawater with different salinity levels from the ocean and storing in separate compartments
- Using ion-exchange membranes that are placed between the compartments containing seawater with different salinity levels to selectively allow the passage of ions, creating a concentration gradient across the membrane
- Generating electric potential difference as the ions move across the membrane
- Reversing the polarity to maintain the process and prevent the membranes from becoming saturated with ions

The principle behind power generation from ocean salinity gradient is based on the Gibbs free energy equation, which states that the free energy available from a system is determined by the enthalpy and entropy changes of the system. The difference in salinity levels between the seawater compartments creates a free energy difference, which can be converted into power. This process can be repeated continuously as long as there is a sufficient difference in salinity levels between the seawater compartments. The two main methods are: (i) pressure retarded osmosis that can convert the osmotic pressure of saline solutions to hydraulic pressure to run the turbine and generate power and (ii) reverse electrodialysis that uses

Figure 1.36 Schematic of salinity gradient power generation with the TaPa-SO$_3$H membrane. Source: Hou [87]/John Wiley & Sons.

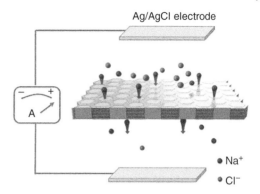

cation- and anion-exchange membranes and different water salinities to transport ions through membranes, such as salt batteries and generate power (Figure 1.36).

This solution has a high potential to generate significant power, especially in areas with a large difference in salinity levels (e.g., coastal areas) between seawater and freshwater. The main drawbacks include: (i) the water requirement that can limit their usability and can impact local ecosystems and (ii) high capital costs, especially for ion-exchange membranes and semi-permeable membranes, and high operational costs for maintaining the concentration gradient across the membranes. The benefits include: (i) compatibility with existing power infrastructure, (ii) secure energy source as it is not subject to supply disruptions, and (iii) sustainable solution since it has minimal environmental impacts and does not release GHG emissions during the operation. Power generation from ocean salinity gradient is still in the early stages of development and has a low TRL. More research and development are needed to improve the efficiency of the process and reduce the costs associated with the technology. Further details about salinity gradient energy have been provided by [84].

1.6.2 Tidal (Two Technologies)

Tidal-based energy is a renewable source from the flow induced by the rise and fall of ocean tides as they ebb and flow due to the gravitational effects of the moon and sun on the oceans, which creates a cyclic rise and fall of water levels in coastal areas. The main tidal power generation technologies are tidal range and tidal stream. Tidal range uses the difference in water levels between high and low tides to generate power. One way of doing this is by building a tidal barrage, which is a dam-like structure that is built across a river estuary or bay. As the tide comes in, water is allowed to flow through turbines in the barrage, generating power. When the tide goes out, the water is released from the barrage back into the estuary or bay. This method requires a large tidal range, the difference in water levels between high and low tide, and is location-specific. Tidal stream uses the kinetic energy of the tidal currents to generate power. Tidal stream turbines are placed underwater and work similarly to wind turbines but are designed to work in water. As the tidal currents flow past the turbines, they rotate and generate power. Tidal stream turbines are placed in areas of strong tidal currents, such as narrow channels or between islands.

The technology readiness of tidal energy generation technologies is at the pre-commercial stage due to turbine design failure, reliability, short lifespan, and accessibility to tidal flows. Tidal stream turbines work similarly to wind turbines but are designed to work in water, and they have a high TRL and are already being used in commercial tidal energy projects in Europe. Tidal barrage is a dam-like structure and has a relatively high TRL, and has been used for power generation in France and the United Kingdom. Dynamic tidal power involves building a long dam-like structure perpendicular to the coastline, with turbines embedded in it to harness the energy of the tides as they flow in and out. It has a relatively high TRL and is being researched and developed in the Netherlands. Tidal kites are a newer technology involving tethered underwater turbines pulled back and forth by tidal currents. The motion of the kites generates electricity, which is transmitted to shore through a cable. Tidal kites are still in the experimental phase, but have shown promising results in pilot projects in Scotland.

Tidal energy was used to operate grain mills over 500 years ago. The principle of power generation from tides involves converting the kinetic energy of the moving water into power. The high density of seawater compared to air can collect energy from tides using smaller turbines than wind turbines. However, these turbines have to be able to collect energy in various directions and survive in harsh marine environments. In both tidal range and tidal stream technologies, turbines are used to capture the energy from the water movement and convert it into power. The turbines are connected to a generator that converts the mechanical energy into electrical energy that can be transmitted to the grid. Tidal power generation is a clean and renewable energy source, as it does not produce GHG emissions. However, it does have some limitations, such as its dependence on the tidal cycle and the availability of suitable locations for tidal power generation. Despite these limitations, tidal power has the potential to play an essential role in meeting our future energy needs. More information has been provided by NREL [88] and EIA [89].

1.6.2.1 Tidal Range (TRL 9)

Tidal range solution harnesses the potential energy stored in different water levels (high and low tides) and converts it to kinetic energy by flowing the water through turbines and generating power. The principle of this solution is similar to hydroelectric power generation by converting water energy stored in a dam to kinetic energy through turbines and generators. Particularly, the tidal barrage acts as a dam, holding back the water during high tide and flowing through turbines during low tide. The tidal barrage is usually across a river estuary or bay and has a high tidal range. The barrage consists of gates to let water enter the tidal basin during high tide and keep it as the tide recedes. The gates are open when the water level in the tidal basin is higher than the sea level, and the water can flow through the turbines to generate power.

The main factors for power output are tidal range size and water flow amount. Recent studies show that a tidal range of at least 3 m (around 10 feet) is needed for economically viable power generation [89]. The power output is highest during spring tides when the tidal range is considerable. The energy output can be predicted and scheduled, as tides are highly predictable and follow a regular cycle. The advantages of tidal range power generation are that it is a clean and renewable energy source that does not release GHG emissions and is highly predictable and reliable. However, the construction of tidal barrages can have environmental impacts, including altering the local ecosystem and affecting fish migration patterns. A two-way tidal range can generate power from both incoming and outgoing tides.

Globally, there are several commercialized tidal barrages with up to 254 MW capacity. The leading companies are SIMEC Atlantis Energy in the United Kingdom and Scotland, Minesto in the United Kingdom and Taiwan, Nova Innovation in the United Kingdom and Scotland, and Tocardo in the Netherlands and Canada. These companies have demonstrated the feasibility and effectiveness of their tidal range technologies for power generation through successful pilot projects and commercial installations.

1.6.2.2 Tidal Stream/Ocean Current (TRL 5)

Tidal stream or ocean current solution harnesses the kinetic energy of moving water to turn the rotor, run mechanical generators, and produce power. This solution uses the principle of rotating turbines, such as axial- or crossflow turbines, reciprocating devices (oscillating hydrofoils), and vortex-induced vibration that are placed in the path of tidal currents or ocean currents. The key factors of this solution include water temperature, depth, density, and salinity, as well as earth rotation and wind. This process is very similar to power generation through wind turbines, and the main difference is that it operates underwater. The main challenge is the harsh marine environment that can damage the turbines.

The main factors for power output include the tidal current strength and turbine size. For instance, narrow channels or between islands can provide strong tidal currents. Tidal stream power generation is a renewable and predictable solution because tides are highly predictable and follow a regular cycle, allowing for accurate power output predictions. This solution is sustainable and has minimal environmental impacts, with no GHG emissions released during power generation. Despite the challenges, several companies have been developing and deploying tidal stream turbines in recent years.

The main parameters of tidal stream power generation are the strength of the tidal currents, the size and number of turbines, and the efficiency of the conversion process. High TRL companies have achieved significant progress in improving the performance of tidal stream turbines in recent years. Tidal stream power generation has a relatively high TRL, with several companies (e.g., Orbital Marine Power, SIMEC Atlantis Energy, and Nova Innovation) developing and deploying commercial-scale projects. Tidal stream power generation is a renewable and clean energy source, but the installation and operation of turbines can have negative environmental impacts, such as noise pollution and the potential for marine life entanglement. Earlier studies show the capital cost of tidal stream power generation is high due to the expensive installation and maintenance of the underwater turbines. The operational cost is low due to no fuel requirements and a long lifespan. More detailed information about ocean currents energy has been provided by [90].

1.7   Case Study: Solar Hybrid Energy Systems

Solar hybrid energy systems use solar collectors with other renewable energy sources for power generation. Since the solar power system is the primary energy source, the science behind this solution is based on converting sunlight into power using PV panels. The benefits include: (i) combining solar energy with other energy sources to ensure a stable and reliable power supply; (ii) using it in remote areas where there is no access to the grid or in areas where there is a high power demand; and (iii) using it as backup power sources for homes, businesses, and industries in case of power outages. The performance of these systems mainly depends on PV panel size, battery capacity, and inverter efficiency. The results of earlier studies show that solar-based energy contribution to annual power generation increased over 10 times during the past decade (Table 1.10).

Solar hybrid energy systems are compatible with other renewable energy sources (e.g., biomass, geothermal, wind, wave, and hydrogen), and can also be integrated with conventional energy sources, such as natural gas or diesel. Examples of solar hybrid energy systems are as follows:

- Solar–biomass: The science behind solar–biomass systems for power generation involves using solar energy to produce heat or power to supplement the energy generated by biomass. Biomass can be any organic matter, such as wood chips, crop residues, or municipal solid waste, that can be converted into energy through processes, such as combustion, pyrolysis, gasification, or

Table 1.10 Relative contribution of resource types in annual capacity additions.

Capacity type	Nameplate capacity additions (GW)											
	2010	2011	2012	2013	2014	2015	2016	2017	2018	2019	2020	2021
Solar	0.9	1.9	3.0	5.7	5.2	6.3	11.4	8.6	8.2	9.7	14.8	18.9
Wind	5.2	6.8	13.0	1.1	4.9	8.6	8.2	7.0	7.6	9.1	17.2	13.4
Storage	0.0	0.0	0.1	0.3	0.0	0.3	0.2	0.2	0.4	0.3	0.7	3.6
Other renewable energy sources	0.4	0.8	1.2	1.5	0.7	0.5	0.5	0.5	0.4	0.3	0.3	0.1
Gas	7.4	12.1	9.8	7.9	9.0	6.6	9.5	12.4	21.8	8.9	7.6	6.2
Coal	6.0	1.7	4.8	1.8	0.1	0.0	0.0	0.0	0.0	0.0	0.0	0.0
Other non-renewable energy sources	1.2	0.8	0.3	0.1	0.2	0.2	1.4	0.1	0.1	0.1	0.0	0.0
Total	21.1	24.2	32.1	18.3	20.1	22.5	31.2	28.9	38.4	28.5	40.7	42.2
Solar % of total	4%	8%	9%	31%	26%	28%	37%	30%	21%	34%	36%	45%

Source: Hitachi, ACP, EIA, Berkeley Lab.

anaerobic digestion. The solar component of the system can be either PV or solar thermal technology. PV systems convert sunlight into electricity, while solar thermal systems use sunlight to heat a fluid, which can then be used to generate steam and power through a steam turbine. In a solar–biomass system, the solar component is used to supplement the energy generated by biomass, thereby reducing the amount of biomass needed to produce the same amount of energy. This can lead to reduced emissions, lower costs, and increased efficiency. Solar–biomass system is one of the most studied solar hybrid systems [91]. Several studies explored solar PV, PVT, concentrated PV (CPV), and concentrated PVT (CPVT) systems and a biomass plant linked with a steam engine or fuel cell to generate and store power and heat for various applications [92]. Also, biomass can be used to produce syngas or biogas to run a gas turbine and generate power and heat for district heating. Earlier studies explored solar–biomass CHP from distillery wastes integrated with a parabolic trough and steam Rankine cycle [93, 94]. One of the main reasons to use the solar–biomass-based CHP is to reduce biomass consumption. The economic analysis results showed that in almost all scenarios, the integrated solar–biomass plant coupled with a parabolic trough solar field and an ORC unit showed positive economic outcomes [95]. Besides, CPV panels use mirrors and lenses to concentrate sunlight on a small area. Concentrated light can increase the efficiency of PV modules. The barriers are the high cost of PV materials, complex manufacturing processes, and the ability to track the sun. Most recent studies evaluated CPVT for both power and heat generation, in which the direct solar radiation and heat on PV cells can be collected with a coolant system. CPVTs have application in residential and commercial buildings, such as hospitals, for space heating, hot water, and air conditioning [96]. Several companies are actively engaged in power generation from solar–biomass systems, such as NextEra Energy, Enel Green Power, and NRG Energy.

- Solar–geothermal: This solution combines the principles of solar and geothermal power generation, using solar energy to heat a fluid (e.g., water) and pump it deep into the earth's crust to access geothermal energy and generate power from hot water or steam. The main benefit is the high efficiency and power output due to combining solar and geothermal energy. The solar–geothermal power generation harnesses the sun's power and the earth's natural heat to generate power. This solution has become increasingly popular as a renewable, sustainable, and efficient method. Earlier studies investigated different solar technologies (e.g., PV, PVT, and CPVT) with geothermal-based multigeneration CHP plants [97, 98]. Their results show that the interest in solar–geothermal hybrid plants is growing due to the rise in domestic hot water demand. Few companies have started to implement this solution on a medium to large scale, such as GreenFire Energy, Sage Geosystems, Eavor, Terrapin Geothermics, and AltaRock Energy.

- Solar–wind: This solution harnesses the sun's and wind's energy to generate power, making it a more reliable and efficient form of renewable energy. Like solar–geothermal, combining solar and wind energy increases the process efficiency and power output. Earlier studies investigated PV–wind hybrid systems, including a fuel cell for hydrogen storage for power and heat generation or stand-alone applications [99, 100]. SinnPower has developed a floating ocean hybrid platform that can generate power from solar, wind, and wave, using 20 kW PV cells, 6 kW wind turbines, and integrated wave converters as an off-grid solution [101]. Recent studies investigated hybrid floating solar PV systems on various water bodies, such as oceans and dams (Figure 1.7). Floating PV systems can be integrated with other energy resources (e.g., hydropower, wind, and wave) to address the world's power demand. Solar–wind energy system is a new technology, and a few companies, such as GE Renewable Energy, Siemens Gamesa, and Vestas, are actively developing and deploying this technology.

- Solar–hydrogen: This solution uses solar energy to generate hydrogen, and store and use it to generate power when needed. The water electrolysis process generates hydrogen using solar energy by splitting water molecules into hydrogen and oxygen, using an electric current from solar PV systems. The stored hydrogen can be used to generate power when no sunlight is available using fuel cells. This solution is a renewable, sustainable, and efficient method for power generation, and the only byproduct is water. However, there are still some challenges to overcome in terms of scaling up the technology and making it cost-effective for widespread use. Earlier studies explored a solar–hydrogen CHP plant, using solar PV array, solar thermal collectors, fuel cells, hydrogen storage tanks, and water storage tanks for remote households [102]. Recent studies explored floating solar farms in the ocean, using PV cells to produce power and fuel, such as methanol or hydrogen from seawater [103]. The main barriers are rigid deployments and energy storage. More detailed information and comparison of solar–hydrogen generation studies, using thin-film methods and different photocatalyst systems (e.g., CdS, TiO_2, and $ZnS–In_2S_3–CuS_2$) have been provided by Gopinath and Nalajala (2021) [104]. Several companies are actively engaged in power generation from solar–hydrogen systems, such as Nel Hydrogen, Siemens Energy, Hydrogenics Corporation, Linde, and Air Liquide.

1.8 Denmark Case Study

One of the world's largest renewable energy-based heating plants is SUNSTORE4 in Denmark, with a significant share of solar-based power and heat [105]. Their main known technologies are solar PV systems and non-concentrated solar thermal systems for residential and district heating needs. Solar–biomass hybrid systems are also common for both small- and large-scale CHP plants. ORC has attracted interest in Denmark for decentralized, small-capacity CHP plants. Due to the increase in biomass use as an energy input in Denmark, the applied energy systems include solar PV arrays, solar flat-plate collectors, biomass burners, heat pumps, and ORC units. Denmark study compared the profitability and environmental benefits of base case (heat supply from natural gas burner) with various cases, such as solar PV or flat-plat collectors with biomass burners or ORC units. The biomass burner operates with wood pellets (as carbon neutral) with high density and low moisture content, and provides heat to the ORC unit when the heating demand is not covered by solar energy. The generated power from ORC units can be used for residential CHP needs. All case studies are based on 100 residential houses with total 490 MWh power demand (average 13.4 kWh/day) and 1,458 MWh heat demand, as well as an average 5.4 kWh/day hot water demand per house for 30 years of plant lifetime. The required solar space (both PV and flat plate) is 1,000 m^2 (10–13 m^2 per house) with 45° tilt angle (faced south). After invertor losses, the PV array can provide up to 158 kWh/m^2 solar energy per year. The flat-plate collectors can provide 424 kWh/m^2 for supplying 75 °C temperature annually. The results indicated that the solar flat-plate collectors with biomass burner plant have the lowest cost annually for heat generation compared to other case studies that can be

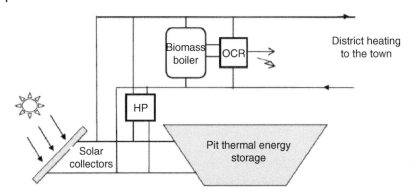

Figure 1.37 Schematic of SUNSTORE configuration in Marstal, Denmark. Source: Adapted from CORDIS EU Research Results [105].

used instead of conventional methods, such as natural gas-based district heating plants. Additionally, the flat-plate collector with biomass burner and ORC unit plant is preferable economically for both heat and power generation, with 9% lower annual cost for 100 houses and 28% lower for 1,000 houses compared with the natural gas solution. The environmental results showed that natural gas and grid power has the highest emission, and those cases with heat pump have the second highest carbon emissions. The cases with biomass solution have the lowest emissions. It is concluded that a large capacity plant from the flat-plate collector with biomass burner and ORC unit plant has significant economic and environmental benefits (Figure 1.37). Earlier studies concluded that small- and large-scale solar–biomass hybrid plants are a sustainable approach, both economically and environmentally, for combined power and heat generation in Nordic climate conditions, such as Denmark [10]. Particularly, they suggested combined ORC with solar thermal collectors and biomass burners. Further details about inputs and outputs, economic and CO_2 emission analyses, and conversion efficiency have been provided by [10].

The latest NREL study compares the efficiency of main PV cells and modules (Figure 1.38). Currently, multijunction cells, especially four-junction or more (concentrator) cells, have the highest efficiency (around 47%), and three-junction (concentrator) cells have the 2nd highest efficiency, around 44%. CIGS and CIGSS modules have the highest efficiencies, with approximately 38.9 and 35.9%, respectively.

These systems are considered mature technologies with high TRLs. They have been widely used in many countries for several years and have proven to be effective in generating power. These systems have high capital costs, but low operational costs. The initial investment in PV panels, batteries, and other components can be expensive, but the operational costs are minimal since solar energy is free

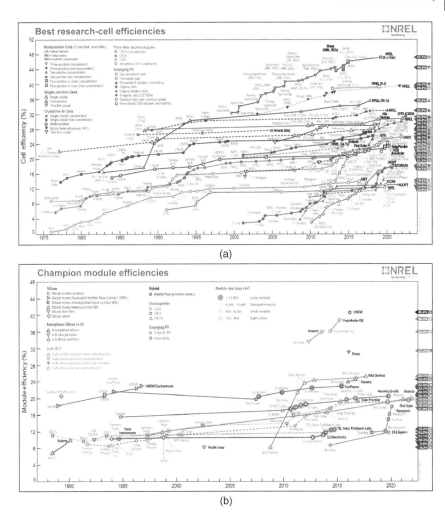

Figure 1.38 Comparison of Solar PV cells (a) and modules (b) efficiency. Source: National Renewable Energy Laboratory (public domain).

and maintenance costs are low. The main benefits include: (i) providing a stable and reliable source of power; (ii) reducing dependence on fossil fuels and contributing to a cleaner environment; (iii) being usable in remote areas where there is no access to the grid; and (iv) serving as backup power sources in case of power outages. The drawbacks include: (i) requiring high capital costs; (ii) some environmental impacts; (iii) requiring adequate sunlight to generate power, which may limit their use in certain areas; and (iv) requiring proper maintenance to ensure optimal performance. Overall, solar hybrid systems are relatively affordable and capable of addressing consumer demands technically and cost-effectively in rural and urban areas. Small-capacity solar hybrid systems are of high interest in remote residential houses.

Self-Check Questions

1. What are the most used materials in PV panels?
2. What are the main techno-economic parameters in solar PV technologies?
3. What are the benefits of flexible solar cells?
4. What is a crystalline silicon (C-Si) solar panel? And how does it work?
5. What are the top three used solar PV panels? And what are their benefits and drawbacks?
6. What are the required materials for C-Si PV?
7. What are the main benefits of multijunction PV cells?
8. What are the main types of organic PV cells?
9. What are the benefits and drawbacks of solar thermal towers for power and heat generation?
10. What are the benefits and drawbacks of parabolic trough systems for power and heat generation?
11. What are the main parameters that affect the performance of linear Fresnel reflector systems?
12. What are the benefits and drawbacks of parabolic trough systems for power and heat generation?
13. What are the main types of power generation from wind?
14. What are the benefits and drawbacks of power generation via onshore wind turbines?
15. What are the benefits and drawbacks of power generation via seabed-fixed offshore wind turbines?
16. What are the benefits and drawbacks of power generation via floating offshore wind turbines?
17. What are the benefits and drawbacks of power generation through hydropower?
18. What are the main geothermal technologies for power generation?
19. What are the benefits and drawbacks of geothermal power generation?
20. What are the benefits and drawbacks of flash steam geothermal power generation?
21. What are the key parameters affecting the performance of binary organic Rankine cycle geothermal power generation?
22. What are the benefits and drawbacks of binary organic Rankine cycle geothermal power generation?

23. What are the benefits and drawbacks of enhanced geothermal systems for power generation?
24. What are the main hydrogen generation methods?
25. What are the main steps for power generation through hydrogen-fired gas turbine systems?
26. What is the operation principle of fuel cells?
27. What are the benefits and drawbacks of hydrogen-fired gas turbines for power generation?
28. What are the benefits and drawbacks of high-temperature fuel cells for power generation?
29. What are the benefits and drawbacks of hybrid fuel cell–gas turbine systems for power generation?
30. What are the major marine technologies for power generation?
31. What are the main wave energy technologies for power generation?
32. What are the main steps during ocean thermal power generation?
33. What are the main types of ocean thermal conversion technologies?
34. What are the basic steps for power generation from ocean salinity gradient?
35. What are the main ocean salinity gradient methods for power generation?
36. What are the benefits and drawbacks of power generation from ocean salinity gradient?
37. What are the main tidal power generation technologies?
38. What are solar hybrid energy systems? And what are their benefits?

References

1 IEA (2021). *Net Zero by 2050: A Roadmap for the Global Energy Sector*.
2 PBL (2020). *Trends in Global CO2 and Total Greenhouse Gas Emissions*; 2020 Report.
3 Twidell, J. (2021). *Renewable Energy Resources*. Taylor & Francis.
4 U.S. EIA (2021). *Independent Statistics and Analysis*. https://www.eia.gov/todayinenergy/detail.php?id=49876.
5 U.S. EIA (2022). *Annual Energy Outlook 2022*.
6 US DOE (2022). *Solar Photovoltaics Supply Chain Review*.
7 Lazard (2022). *Levelized Cost of Energy and of Storage 2020*.
8 Khan, J. and Arsalan, M.H. (2016). Solar power technologies for sustainable electricity generation–a review. *Renewable and Sustainable Energy Reviews* 55: 414–425.
9 US DOE (2021). *Solar Futures Study*.

10 Modi, A., Bühler, F., Andreasen, J.G., and Haglind, F. (2017). A review of solar energy based heat and power generation systems. *Renewable and Sustainable Energy Reviews* 67: 1047–1064.

11 U.S. DOE (2021). *Photovoltaics Progress and Goals*. https://www.energy.gov/eere/solar/photovoltaics.

12 Franco A, Fantozzi F. Experimental analysis of a self consumption strategy for residential building: the integration of PV system and geothermal heat pump. *Renewable Energy* 2016;86:1075–1085. doi:https://doi.org/10.1016/j.renene.2015.09.030.

13 Ahmadi, P., Dincer, I., and Rosen, M.A. (2015). Transient thermal performance assessment of a hybrid solar-fuel cell system in Toronto, Canada. *International Journal of Hydrogen Energy* 40: 7846–7854.

14 Shabani, B. and Andrews, J. (2011). An experimental investigation of a PEM fuel cell to supply both heat and power in a solar-hydrogen RAPS system. *International Journal of Hydrogen Energy* 36: 5442–5452.

15 Zafar, S. and Dincer, I. (2014). Thermodynamic analysis of a combined PV/T–fuel cell system for power, heat, fresh water and hydrogen production. *International Journal of Hydrogen Energy* 39: 9962–9972.

16 Ramanujam, J., Bishop, D.M., Todorov, T.K. et al. (2020). Flexible CIGS, CdTe and a-Si: H based thin film solar cells: a review. *Progress in Materials Science* 110: 100619.

17 Tawalbeh, M., Al-Othman, A., Kafiah, F. et al. (2021). Environmental impacts of solar photovoltaic systems: a critical review of recent progress and future outlook. *Science of The Total Environment* 759: 143528.

18 Nassiri Nazif K, Daus A, Hong J, Lee N, Vaziri S, Kumar A, et al. High-specific-power flexible transition metal dichalcogenide solar cells. *Nature Communications* 2021;12:7034. doi:https://doi.org/10.1038/s41467-021-27195-7.

19 Forbes (2021). *How Much Do Solar Panels Cost?*

20 Liang TS, Pravettoni M, Deline C, Stein JS, Kopecek R, Singh JP, et al. A review of crystalline silicon bifacial photovoltaic performance characterisation and simulation. *Energy and Environmental Science* 2019;12:116–148. doi:https://doi.org/10.1039/C8EE02184H.

21 US DOE (2022). *Multijunction III–V Photovoltaics Research*.

22 Liu, Z., Lv, H., Hu, Y. et al. (2021). A review of modeling of luminescent coupling effect in multi-junction solar cell based on diode equation. *International Journal of Low-Carbon Technologies* 16: 1519–1528.

23 Kerestes C, Polly S, Forbes D, Bailey C, Podell A, Spann J, et al. Fabrication and analysis of multijunction solar cells with a quantum dot (In)GaAs junction. *Progress in Photovoltaics: Research and Applications* 2014;22:1172–1179. doi:https://doi.org/10.1002/pip.2378.

24 Gadzanku, S., Lee, N., and Dyreson, A. (2022). *Enabling Floating Solar Photovoltaic (FPV) Deployment: Exploring the Operational Benefits of Floating Solar-Hydropower Hybrids.* Golden, CO (United States): National Renewable Energy Laboratory (NREL).

25 Solomin E, Sirotkin E, Cuce E, Selvanathan SP, Kumarasamy S. Hybrid floating solar plant designs: a review. *Energies* 2021;14:2751. doi:https://doi.org/10.3390/en14102751.

26 US DOE (2022). *Cadmium Telluride.*

27 ASCA (2022). *The Photovoltaic Solution that Unlocks your Imagination.*

28 Kettle J, Aghaei M, Ahmad S, Fairbrother A, Irvine S, Jacobsson JJ, et al. Review of technology specific degradation in crystalline silicon, cadmium telluride, copper indium gallium selenide, dye sensitised, organic and perovskite solar cells in photovoltaic modules: understanding how reliability improvements in mature technologies can enhance emerging technologies. *Progress in Photovoltaics: Research and Applications* 2022;30:1365–1392. doi:https://doi.org/10.1002/pip.3577.

29 US DOE (2022). *Copper Indium Gallium Diselenide.*

30 Mavlonov, A., Razykov, T., Raziq, F. et al. (2020). A review of Sb2Se3 photovoltaic absorber materials and thin-film solar cells. *Solar Energy* 201: 227–246.

31 Wang Y, Zhan X. Layer-by-layer processed organic solar cells. *Advanced Energy Materials* 2016;6:1600414. doi:https://doi.org/10.1002/aenm.201600414.

32 Kong J, Nordlund D, Jin JS, Kim SY, Jin S-M, Huang D, et al. Underwater organic solar cells via selective removal of electron acceptors near the top electrode. *ACS Energy Letters* 2019;4:1034–1041. doi:https://doi.org/10.1021/acsenergylett.9b00274.

33 US DOE (2022). *Organic Photovoltaics Research.*

34 Epishine (2022). *Epishine is the Leading Developer and Manufacturer of Printed Organic Solar Cells.*

35 US DOE (2022). *Perovskite Solar Cells.*

36 GCL (2022). *Perovskite Triggers PV Technology Revolution. Industry Big Shots Gather in SuzhouCorporate News.*

37 Norwood, Z. and Kammen, D. (2012). Life cycle analysis of distributed concentrating solar combined heat and power: economics, global warming potential and water. *Environmental Research Letters* 7: 044016.

38 Desideri U, Campana PE. Analysis and comparison between a concentrating solar and a photovoltaic power plant. *Applied Energy* 2014;113:422–433. doi:https://doi.org/10.1016/j.apenergy.2013.07.046.

39 NREL (2022). *Life Cycle Greenhouse Gas Emissions from Concentrating Solar Power.*

40 NREL (2022). *Life Cycle Greenhouse Gas Emissions from Solar Photovoltaics.*

41 Desideri U, Zepparelli F, Morettini V, Garroni E. Comparative analysis of concentrating solar power and photovoltaic technologies: technical and environmental evaluations. *Applied Energy* 2013;102:765–784. doi:https://doi.org/10.1016/j.apenergy.2012.08.033.

42 Union of Concerned Scientists. Environmental impacts of solar power. 2013. https://www.ucsusa.org/resources/environmental-impacts-solar-power ().

43 Ho CK. Computational fluid dynamics for concentrating solar power systems. *WIREs Energy and Environment* 2014;3:290–300. doi:https://doi.org/10.1002/wene.90.

44 US EIA (2022). *Solar Thermal Power Plants.*

45 Abed N, Afgan I. An extensive review of various technologies for enhancing the thermal and optical performances of parabolic trough collectors. *International Journal of Energy Research* 2020;44:5117–5164. doi:https://doi.org/10.1002/er.5271.

46 Crema, L., Alberti, F., Wackelgard, E. et al. (2014). Novel system for distributed energy generation from a small scale concentrated solar power. *Energy Procedia* 57: 447–456.

47 Abdelhady S, Borello D, Tortora E. Design of a small scale stand-alone solar thermal co-generation plant for an isolated region in Egypt. *Energy Conversion and Management* 2014;88:872–882. doi:https://doi.org/10.1016/j.enconman.2014.08.066.

48 Borunda M, Jaramillo OA, Dorantes R, Reyes A. Organic Rankine cycle coupling with a parabolic trough solar power plant for cogeneration and industrial processes. *Renewable Energy* 2016;86:651–663. doi:https://doi.org/10.1016/j.renene.2015.08.041.

49 Sun J, Zhang Z, Wang L, Zhang Z, Wei J. Comprehensive review of line-focus concentrating solar thermal technologies: parabolic trough collector (PTC) vs linear fresnel reflector (LFR). *Journal of Thermal Science* 2020;29:1097–1124. doi:https://doi.org/10.1007/s11630-020-1365-4.

50 US DOE (2022). *History of U.S. Wind Energy.*

51 US DOE (2022). *How Do Wind Turbines Work?*

52 US DOE (2022). *Land-Based Wind Market Report.*

53 US DOE (2022). *Wind Turbines: The Bigger, The Better.*

54 US DOE (2022). *Top 10 Things You Didn't Know About Wind Power.*

55 US DOE (2022). *Offshore Wind Market Report.*

56 Jansen M, Staffell I, Kitzing L, Quoilin S, Wiggelinkhuizen E, Bulder B, et al. Offshore wind competitiveness in mature markets without subsidy. *Nature Energy* 2020;5:614–622. doi:https://doi.org/10.1038/s41560-020-0661-2.

57 Schmidt, H., de Vries, G., Renes, R.J., and Schmehl, R. (2022). The social acceptance of airborne wind energy: a literature review. *Energies* 15: 1384.

58 McKenna, R., Pfenninger, S., Heinrichs, H. et al. (2022). High-resolution large-scale onshore wind energy assessments: a review of potential definitions, methodologies and future research needs. *Renewable Energy* 182: 659–684.

59 Weber J, Marquis M, Cooperman A, Draxl C, Hammond R, Jonkman J, et al. *Airborne Wind Energy*. National Renewable Energy Lab. (NREL), Golden, CO (United States); 2021. doi:https://doi.org/10.2172/1813974.

60 US DOE (2021). *New Report Discusses Opportunities and Challenges for Airborne Wind Energy.*

61 US DOE (2022). *Types of Hydropower Turbines.*

62 US DOE (2022). *Hydropower Program.*

63 US DOE (2022). *Pumped Storage Hydropower: A Key Part of Our Clean Energy Future.*

64 US DOE (2022). *Environmental and Hydrologic Systems Science.*

65 Oak Ridge National Laboratory (2022). *Hydrosource.*

66 Jin Y, Behrens P, Tukker A, Scherer L. Water use of electricity technologies: a global meta-analysis. *Renewable and Sustainable Energy Reviews* 2019;115:109391. doi:https://doi.org/10.1016/j.rser.2019.109391.

67 US DOE (2021). *Hydropower Market Reports.*

68 Darwdown (2022). *Project Drawdown.* Darwdown.

69 US DOE (2022). *Geothermal Technologies Office.*

70 Robins J, Kolker A, Flores-Espino F, Pettitt W, Schmidt B, Beckers K, et al. *U.S. Geothermal Power Production and District Heating Market.* 2021. doi:https://doi.org/10.2172/1808679.

71 Buonocore E, Vanoli L, Carotenuto A, Ulgiati S. Integrating life cycle assessment and energy synthesis for the evaluation of a dry steam geothermal power plant in Italy. *Energy* 2015;86:476–487. doi:https://doi.org/10.1016/j.energy.2015.04.048.

72 Loni R, Mahian O, Najafi G, Sahin AZ, Rajaee F, Kasaeian A, et al. A critical review of power generation using geothermal-driven organic Rankine cycle. *Thermal Science and Engineering Progress* 2021;25:101028. doi:https://doi.org/10.1016/j.tsep.2021.101028.

73 U.S. DOE (2022). *Organic Rankine Cycle (ORC) System Basics.* https://chp.ecatalog.ornl.gov/benefits/orc-basics.

74 Soltani M, Nabat MH, Razmi AR, Dusseault MB, Nathwani J. A comparative study between ORC and Kalina based waste heat recovery cycles applied to a green compressed air energy storage (CAES) system. *Energy Conversion and Management* 2020;222:113203. doi:https://doi.org/10.1016/j.enconman.2020.113203.

75 NREL (2022). *Sedimentary and Enhanced Geothermal Systems.*

76 US DOE (2022). *How Wind Energy Can Help Clean Hydrogen Contribute to a Zero-Carbon Future.*

77 Kawasaki Heavy Industries (2022). *Toward the Realization of a Hydrogen Society.* Kawasaki Heavy Industries.

78 Yue, M., Lambert, H., Pahon, E. et al. (2021). Hydrogen energy systems: a critical review of technologies, applications, trends and challenges. *Renewable and Sustainable Energy Reviews* 146: 111180.

79 Koç, Y., Yağlı, H., Görgülü, A., and Koc, A. (2020). Analysing the performance, fuel cost and emission parameters of the 50 MW simple and recuperative gas turbine cycles using natural gas and hydrogen as fuel. *International Journal of Hydrogen Energy* 45: 22138–22147.

80 Haider R, Wen Y, Ma Z-F, Wilkinson DP, Zhang L, Yuan X, et al. High temperature proton exchange membrane fuel cells: progress in advanced materials and key technologies. *Chemical Society Reviews* 2021;50:1138–1187. doi:https://doi.org/10.1039/D0CS00296H.

81 Buonomano A, Calise F, d'Accadia MD, Palombo A, Vicidomini M. Hybrid solid oxide fuel cells–gas turbine systems for combined heat and power: a review. *Applied Energy* 2015;156:32–85. doi:https://doi.org/10.1016/j.apenergy .2015.06.027.

82 National Renewable Energy Laboratory (2022). *Ocean Energy? River Power? There Is a Toolkit for That.* https://www.nrel.gov/news/program/2022/marine-hydrokinetic-toolkit-update.html.

83 U.S. DOE (2022). *Marine and Hydrokinetic Technology.* https://www.energy .gov/eere/water/marine-energy-glossary.

84 OpenEI (2022). *Wave Energy.*

85 Ascari MB, Hanson HP, Rauchenstein L, Van Zwieten J, Bharathan D, Heimiller D, et al. (2012). Ocean thermal extractable energy visualization.

86 OpenEI (2022). *Ocean Thermal Energy Conversion.*

87 Hou S, Ji W, Chen J, Teng Y, Wen L, Jiang L. Free-standing covalent organic framework membrane for high-efficiency salinity gradient energy conversion. *Angewandte Chemie International Edition* 2021;60:9925–9930. doi:https://doi .org/10.1002/anie.202100205.

88 National Renewable Energy Laboratory (2021). *Turning the Tide for Renewables in Alaska.*

89 EIA (2022). *Tidal Power.*

90 OpenEI (2022). *Current Energy.*

91 Burin EK, Buranello L, Giudice PL, Vogel T, Görner K, Bazzo E. Boosting power output of a sugarcane bagasse cogeneration plant using parabolic trough collectors in a feedwater heating scheme. *Applied Energy* 2015;154:232–241. doi:https://doi.org/10.1016/j.apenergy.2015.04.100.

92 Hosseini M, Dincer I, Rosen MA. Investigation of a hybrid photovoltaic-biomass system with energy storage options. *Journal of Solar Energy Engineering* 2014;136 034504. doi:https://doi.org/10.1115/1.4026637.

93 Kaushika ND, Mishra A, Chakravarty MN. Thermal analysis of solar biomass hybrid co-generation plants. *International Journal of Sustainable Energy* 2005;24:175–186. doi:https://doi.org/10.1080/14786450500291909.

94 Sterrer R, Schidler S, Schwandt O, Franz P, Hammerschmid A. Theoretical analysis of the combination of CSP with a biomass CHP-plant using ORC-technology in Central Europe. *Energy Procedia* 2014;49:1218–1227. doi:https://doi.org/10.1016/j.egypro.2014.03.131.

95 Karellas S, Braimakis K. Energy–exergy analysis and economic investigation of a cogeneration and trigeneration ORC–VCC hybrid system utilizing biomass fuel and solar power. *Energy Conversion and Management* 2016;107:103–113. doi:https://doi.org/10.1016/j.enconman.2015.06.080.

96 Papadopoulos, A., Tsoutsos, T., Frangou, M. et al. (2017). Innovative optics for concentrating photovoltaic/thermal (CPVT) systems–the case of the PROTEAS Solar Polygeneration System. *International Journal of Sustainable Energy* 36: 775–786.

97 Calise F, Cipollina A, Dentice d'Accadia M, Piacentino A. A novel renewable polygeneration system for a small Mediterranean volcanic island for the combined production of energy and water: dynamic simulation and economic assessment. *Applied Energy* 2014;135:675–693. doi:https://doi.org/10.1016/j .apenergy.2014.03.064.

98 Bicer Y, Dincer I. Analysis and performance evaluation of a renewable energy based multigeneration system. *Energy* 2016;94:623–632. doi:https://doi.org/10 .1016/j.energy.2015.10.142.

99 Lacko R, Drobnič B, Mori M, Sekavčnik M, Vidmar M. Stand-alone renewable combined heat and power system with hydrogen technologies for household application. *Energy* 2014;77:164–170. doi:https://doi.org/10 .1016/j.energy.2014.05.110.

100 Ozlu S, Dincer I. Development and analysis of a solar and wind energy based multigeneration system. *Solar Energy* 2015;122:1279–1295. doi:https://doi.org/ 10.1016/j.solener.2015.10.035.

101 SINNPower (2022). *Floating Ocean Hybrid Platform*.

102 Assaf J, Shabani B. Transient simulation modelling and energy performance of a standalone solar-hydrogen combined heat and power system integrated with solar-thermal collectors. *Applied Energy* 2016;178:66–77. doi:https://doi .org/10.1016/j.apenergy.2016.06.027.

103 Patterson BD, Mo F, Borgschulte A, Hillestad M, Joos F, Kristiansen T, et al. Renewable CO_2 recycling and synthetic fuel production in a marine environment. *Proceedings of the National Academy of Sciences of the United*

States of America 2019;116:12212–12219. doi:https://doi.org/10.1073/pnas .1902335116.

104 Gopinath, C.S. and Nalajala, N. (2021). A scalable and thin film approach for solar hydrogen generation: a review on enhanced photocatalytic water splitting. *Journal of Materials Chemistry A* 9: 1353–1371.

105 CORDIS EU Research Results (2019). Innovative, multi-applicable-cost efficient hybrid solar (55%)and biomass energy (45%) large scale (district) heating system with long term heat storage and organic Rankine cycle electricity production | SUNSTORE 4 Project | Fact Sheet | FP7 | CORDIS | European Commission. https://cordis.europa.eu/project/id/249800.

2

Power Generation (Low-Carbon Solutions)

According to the U.S. Energy Information Administration (EIA), residential, commercial, and industrial were the major power consumers and end-use sectors in 2021 in the United States, with using over 1,500, 1,300, and 1,000 billion kilowatt-hours of power, respectively [1]. Currently, natural gas is the primary source of electricity generation in the United States, and renewables accounted for around 20% in 2021 and are expected to increase to 42% by 2050 (Table 2.1).

This chapter provides 17 low-emission technologies with high technology readiness levels, along with a case study about carbon capture, utilization, and sequestration (CCUS) hybrid energy systems. Table 2.2 presents the U.S. renewable energy consumption, using the EIA forecasting tool.

2.1 Nuclear

Nuclear-based energy comes from splitting uranium atoms in a process called fission to heat water into steam for powering steam turbines that can generate power (Figure 2.1). Nuclear energy is a reliable, low-emission energy source that can potentially address global GHG emissions from fossil fuel-based energy in the short term [2]. The first nuclear power plants for large-scale energy generation came online in the mid-1950s. The key parameters to control nuclear energy production are temperature for reaction rate, water (the medium) for neutron absorber rate, and control rods for neutron availability to the fissile materials. The top 5 countries with high nuclear energy production are the United States, France, China, Russia, and South Korea, with over 30%, 14%, 12%, 7%, and 5% global share, respectively. In the United States, nuclear-based energy has contributed around 20% of power generation over the past 20 years, which is the largest non-CO_2-emitting power generation technology with over 60 years of operational lifetime. The main challenge is cost reduction to improve competitiveness compared to other low-carbon energy sources. The wastes (e.g., uranium

Net-Zero and Low Carbon Solutions for the Energy Sector: A Guide to Decarbonization Technologies, First Edition. Amin Mirkouei.
© 2024 John Wiley & Sons, Inc. Published 2024 by John Wiley & Sons, Inc.

Table 2.1 Share of the U.S. power generation.

Source	2021	2022[a]	2023[a]	2050[a]
Natural gas (%)	37	38	36	34
Coal (%)	23	20	19	10
Renewables (%)	20	22	24	42
Nuclear (%)	20	19	20	12
U.S. CO_2 emissions (billion metric tons)	4.90	4.98	4.84	–

a) Forecasted.
Source: Adapted from U.S. EIA [1].

Table 2.2 U.S. Renewable energy consumption (Quadrillion Btu).

Source	2021	2022[a]	2023[a]
Renewable diesel and biodiesel	0.386	0.464	0.556
Biofuel losses and coproducts	0.789	0.812	0.796
Ethanol	1.147	1.149	1.149
Geothermal	0.206	0.207	0.206
Hydropower[b]	2.283	2.390	2.501
Solar	1.519	1.889	2.360
Waste biomass	0.431	0.419	0.416
Wood biomass	2.087	2.084	2.145
Wind	3.372	3.900	4.129

a) Forecasted.
b) Hydropower generated by pumped storage is not included in renewable energy.
Source: Adapted from U.S. EIA [1].

and tritiated water) from nuclear power plants can be used in thermoelectric heating and energy generation. Since thorium (Th) is three times more abundant than uranium (U) and has radioactive characteristics, it can be used in nuclear reactors. Thorium absorbs neutrons through the fission process and becomes a new element, such as uranium 233 that can be used in nuclear reactors. Also, thorium contains less radioactive materials compared to uranium, so waste disposal should be easier [3].

The global nuclear power capacity was over 390 GW, using 442 reactors in 32 countries by the end of 2020, around a third of global low-carbon power [2].

Figure 2.1 Example of advanced light-water nuclear technology for power generation. Source: Wiley Digital Library.

Nuclear energy can reduce around 2 gigatons of GHG emissions per year and cut over 55 gigatons of GHG emissions over the last 50 years. International Atomic Energy Agency (IAEA) studies show that nuclear power will reach over 710 GW by 2050, supplying 11% of global power [2]. Beyond power generation, nuclear energy can support hydrogen production [4], district heating, desalination, and industrial process heat applications. However, the interest in nuclear power plants has been reduced due to inherent risks, high prices, radioactive waste, and disasters, such as Fukushima in 2011 and Chernobyl in 1986. Figure 2.2 presents

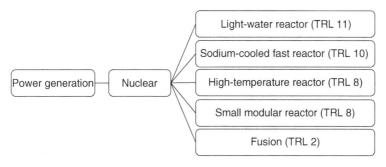

Figure 2.2 Main nuclear technologies for power generation.

the leading nuclear power generation technologies. High-temperature gas-cooled reactors can provide both power and heat for different industrial uses. The versatile test reactor can test new materials and reactor types to generate 10 times more energy and reduce nuclear waste [5]. In the United States and Canada, the latest studies are exploring new molten salt reactor (Terrestrial Energy), the first small modular reactor (NuScale Power) in Oregon, and new fusion reactors by General fusion in Canada, Helion Energy in Redmond, and CTFusion in Seattle.

2.1.1   Light-Water Reactor (TRL 11)

Light-water nuclear reactors operate by using a controlled nuclear reaction to generate heat, which is then used to create steam to power turbines and generate power. The basic science behind nuclear reactions is well understood, and the design of light-water reactors has been refined over decades of research and development. Around 90% of current operational nuclear power capacity uses light water, and the vast majority of under-construction reactors will be cooled with light water due to the high power outputs between 1,000 and 1,700 MW (megawatt) capacity [6]. High-temperature and high-pressure (supercritical) water-cooled reactors have different varieties, both open or closed fuel cycle or thermal or fast neutron spectra with outlet temperature between 510 and 625 °C and various sizes, i.e., small (300–700 MWe) and large (1,000–1,500 MWe). The molten-salt reactor is a closed fuel cycle (Figure 2.3), and fluoride salts cooled reactor in any neutron spectrum (thermal with a thorium fuel cycle or fast with a uranium–plutonium fuel cycle), with outlet temperature between 700 and 800 °C and capacity around 1,000 MWe [7].

The benefits of light-water reactors include: (i) flexibility and compatibility with other power generation and transmission systems; (ii) high efficiency with minimal interruptions; and (iii) near-zero carbon energy source. The challenges are: (i) strict security protocols to prevent unauthorized access to nuclear materials, sabotage, and other security threats; (ii) vulnerability to weather or natural disasters; (iii) nuclear materials and waste management, which are radioactive and require careful handling and disposal; and (iv) environmental impacts on aquatic ecosystems, as they require large amounts of water for cooling. These reactors are designed to be highly functional, generating large amounts of power with relatively low operating costs. However, they require significant maintenance and regulatory oversight to ensure safe and reliable operation. Light-water nuclear reactors are relatively easy to operate and maintain but require specialized knowledge and training. Additionally, handling nuclear materials and waste requires careful attention to safety and regulatory compliance.

The techno-economic analysis shows that the capital costs of building these reactors are high due to the complex and expensive equipment and deployment

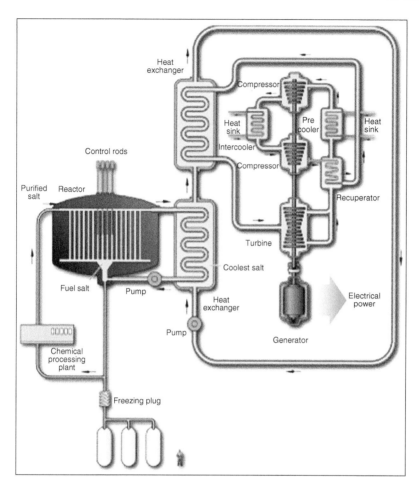

Figure 2.3 Schematic of the molten salt reactor for power generation. Source: U.S. Department of Energy, Office of Nuclear Energy (public domain) [8].

barrier. The operating cost is lower than other forms of power generation, as nuclear fuel can generate a high amount of power. The security of nuclear materials and facilities is a critical concern for the safe operation of nuclear power plants. Major accidents can release radioactive materials into the environment and cause long-term environmental damage. Power generation by light-water reactors is a mature solution with decades of operational experience and well-established infrastructure. The latest studies mainly focus on improving safety and addressing concerns related to nuclear waste management.

2.1.2 Sodium-Cooled Fast Reactor (TRL 10)

Sodium-cooled fast reactors operate on the principle of nuclear fission, which generates heat that is used to produce steam and generate power. They are usually closed fuel cycle and fast neutron spectrum reactors, using liquid sodium as a coolant with an outlet temperature between 500 and 550 °C (Figure 2.4) [8]. The reactor can be arranged in a pool layout with different fuel options (e.g., mixed oxide and mixed metal alloy) and sizes, such as small (50–150 MWe) or large (600–1,500 MWe). Several countries have experience using sodium-cooled fast reactors. For example, Russia has two operational sodium-cooled reactors (BN-600 since 1980 and BN-800 since 2016), and a new reactor is under

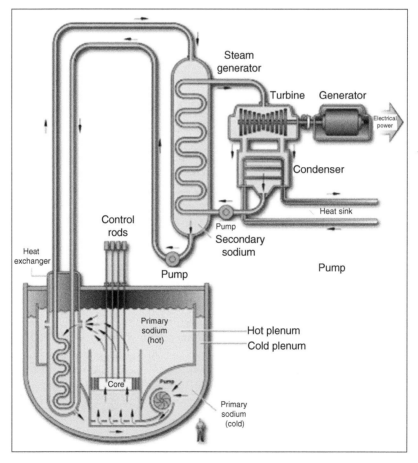

Figure 2.4 Schematic of sodium-cooled fast reactor for power generation. Source: U.S. Department of Energy, Office of Nuclear Energy (public domain) [8].

construction (BN-1200). Also, China has had sodium-cooled reactors since 2010, generating around 20 MW.

Sodium-cooled fast reactors can be used similarly to light-water reactors to generate power, but they require specialized infrastructure and transmission systems due to their unique cooling system. They also need strict security protocols to prevent unauthorized access to nuclear materials and to prevent sabotage or other security threats. Additionally, using liquid sodium as the coolant poses additional challenges for safety and security. They are designed to operate at high levels of efficiency, with minimal downtime or interruptions. However, their performance can be impacted by external factors, such as natural disasters. Sodium-cooled reactors have the potential to be a sustainable energy source due to their ability to use nuclear waste as fuel. The production of nuclear fuel and the management of nuclear waste present sustainability challenges that must be addressed.

Sodium-cooled fast reactors have been under development for several decades, and are designed to be highly functional, generating large amounts of power with relatively low operating costs. However, they are more complex than light-water reactors and require more advanced technology and maintenance. These reactors are more challenging to operate and maintain than light-water reactors due to the highly corrosive nature of liquid sodium and the need for specialized training and expertise. Sodium-cooled fast reactors have high power density and low coolant volume fraction to reduce waste and increase yield. The TRL of sodium-cooled fast reactors is lower compared to light-water reactors, which requires additional research and development to improve safety, efficiency, and sustainability before they can be widely deployed for power generation.

The environmental impacts of sodium-cooled fast reactors are: (i) nuclear waste generation, which is radioactive and requires careful handling and disposal; (ii) significant impacts on aquatic ecosystems, as they require large amounts of water for cooling; and (iii) high risk due to using radioactive materials that can cause long-term environmental damage. The techno-economic analysis shows that the capital costs of building a sodium-cooled fast reactor are typically high, due to the complex and expensive equipment required. However, the operating costs of running sodium-cooled fast reactors are often lower than other low-carbon power generation processes. Thus, the advantages of these reactors include low operation costs and low carbon emissions. The disadvantages include high risk and high capital cost. The gas-cooled fast reactor has the advantages of a fast neutron core and helium coolant with 850 °C outlet temperature, a closed fuel cycle, and a capacity of up to 1,200 MWe. The lead-cooled fast reactor is a closed fuel cycle and fast neutron spectrum reactor with outlet temperature between 480 and 570 °C and different sizes, i.e., small (20–180 MWe) and large (600–1,000 MWe). More detailed information about sodium-cooled fast reactors by the U.S. Department of Energy and Idaho National Laboratory can be found at [9].

2.1.3 High-Temperature Reactor (TRL 8)

The science behind high-temperature nuclear reactors is based on nuclear fission, generating heat that can be used in combined energy systems for heat and power generation (Figure 2.5) [7]. These reactors are open fuel cycle, helium-cooled, and thermal neutron spectrum reactors with outlet temperatures

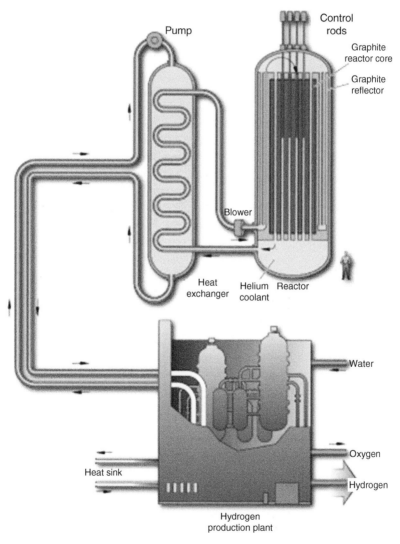

Figure 2.5 Schematic of a high-temperature reactor for power generation. Source: U.S. Department of Energy, Office of Nuclear Energy (public domain) [8].

between 900 and 1,000 °C and capacity between 250 and 300 MWe. High- and very high-temperature reactors have been under development for several decades and have lower TRL compared to traditional reactors.

The benefits of high-temperature nuclear reactors include: (i) highly functional for generating a high amount of power with low operating costs and minimal downtime; (ii) higher efficiency than other nuclear reactors due to high- and very high-temperature reactors that can produce both power and heat for several applications, such as water desalination or hydrogen production; (iii) flexibility and compatibility with other power and heat generation systems; and near-zero carbon energy source. The challenges include: (i) special training requirements for both running the reactors and handling nuclear materials and waste, (ii) security concerns and environmental impacts due to radioactive materials, and (iii) high-temperature requirements and deployment barriers. The environmental impacts of high-temperature reactors are similar to those of other nuclear reactors, such as nuclear waste, storage and disposal barriers, and large water requirements for cooling.

2.1.4 Small Modular Reactor (TRL 8)

Small modular reactors operate on the nuclear fission principle, using energy from controlled nuclear chain reactions to generate heat and power. Currently, the U.S. Department of Energy supports developing advanced small modular reactors using light-water as a coolant (Figure 2.6) [10]. They can be used in remote locations or countries with smaller grids using nonlight water coolants, such as helium, liquid metal, or liquid salt. Russia's first advanced small modular reactor started power generation commercially in 2020 with around 70 MW capacity [2].

The benefits of small modular reactors include: (i) lower capital costs compared to other reactors; (ii) highly functional, modular, and scalable, allowing them to be easily deployed and scaled up as needed; (iii) portable and small physical foot-prints; (iv) high compatibility, flexibility, and efficiency compared to traditional reactors; and (v) zero or very low carbon emissions. The challenges include: (i) security and environmental concerns; (ii) high vulnerability due to their small size; and (iii) sustainability challenges, including nuclear materials and waste management. This solution is easy to operate and maintain with lower staffing requirements than traditional nuclear reactors. These reactors have been under development for several decades, and they vary in size and capacity between 10 and 100 MW, using light water, gases, liquid metals, or salts as coolants. They have high TRL, with several designs currently undergoing regulatory review and approval. Additionally, some small reactors use alternative materials or elements (e.g., thorium) that are more abundant and potentially less expensive than traditional nuclear materials.

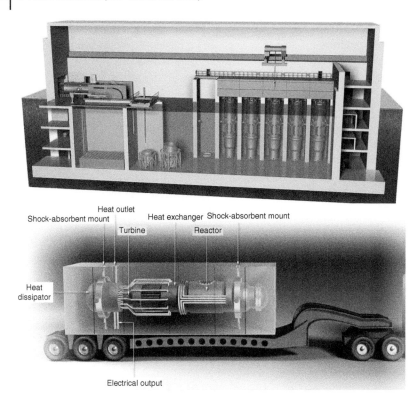

Figure 2.6 Schematic of the advanced small modular reactor (top) and microreactor (bottom). Source: U.S. Department of Energy and Idaho National Laboratory (public domain) [10].

The environmental impacts of small modular reactors are: (i) radioactive nuclear waste generation, which requires careful handling and disposal; (ii) impacts on ecosystems due to the water requirement for cooling; and (iii) potential risks due to the release of radioactive materials. The capital cost of developing a small modular power plant is lower than that of the traditional large-scale power plants due to the simplified design and size, about 1/10 to 1/4 of the size of large-scale plants [11]. Small reactors can be more efficient than large ones, but the operational cost of running the small reactors is higher than that of the larger reactors due to economies of scale. Small modular reactors have the potential to be more efficient and cost-effective than other types of nuclear reactors due to their simplified design and the potential for cogeneration of power and heat. The main disadvantage compared to larger-scale power generation is the low overall power output. More detailed information has been provided by the U.S. Department of Energy, Office of Nuclear Energy [12].

2.1.5 Fusion (TRL 2)

Fission and fusion processes can produce massive energy through nuclear reactions than other sources from splitting larger atoms (fission) or joining smaller atoms (fusion) [13]. The fission process has been used mainly to heat water into steam for powering steam turbines that can generate power by splitting larger atoms into smaller ones. Currently, uranium and plutonium are mainly used for fission processes. However, fusion process can generate energy when two lighter atoms slam together and form a heavier atom, such as two hydrogen atoms to a helium atom, similar to the sun. While the basic science behind nuclear fusion is well understood, developing a practical and scalable fusion reactor remains a significant scientific and engineering challenge (Figure 2.7). The latest research and

Figure 2.7 Schematic of a fusion reactor. Source: U.S. Department of Energy (public domain) [14].

development are in France to build an International Thermonuclear Experimental Reactor that can generate 15 MW thermal power in 5 s. The first operation is scheduled for the end of 2025. Fusion does not generate highly radioactive wastes; however, it is difficult to run it for a longer time due to the high amount of pressure and temperature needed to form nuclei together [13].

Fusion reactors harness the energy released by nuclear fusion reactions that can be a highly functional and efficient energy source with high power output and low waste. The benefits include: (i) lower security risks, (ii) lower radiation risks compared to traditional nuclear reactors as they do not use fissile materials (e.g., uranium or plutonium), (iii) flexibility and compatibility with other power generation and transmission systems, (iv) sustainable method and energy source since it uses hydrogen as fuel and generate no carbon emissions or radioactive wastes, (v) cost-competitive energy source in the long term due to low fuel costs, using hydrogen as fuel that is abundant and easily accessible, and (vi) lower environmental impacts and waste compared to traditional methods.

The drawbacks include: (i) environmental impacts due to high land use and energy requirements, (ii) high capital and operational costs due to the complexity and other requirements (very high temperature and pressure), and (iii) difficulty in handling and maintaining due to the extreme conditions inside a fusion reactor. There is limited techno-economic data on their feasibility. Earlier studies estimated that the costs and benefits of fusion power generation would be higher than fission reactors or fossil fuels. Fusion nuclear reactors are in the experimental stage and have low TRLs due to process complexity and requirements. Overall, fusion reactors may become more cost-competitive as the technology matures. Further details about this solution have been provided by the U.S. Department of Energy, Office of Science [14].

2.2 Coal

Power generation from coal mainly involves four main steps: coal combustion, steam generation, cooling, and emission control. During the combustion process, coal is pulverized into fine powders and fed into a boiler to generate high-temperature and high-pressure steam that can be used to run steam turbines and power generators. Steam turbines convert the thermal energy of steam to mechanical energy and drive the generators to produce power. The cooling process uses cooling towers or condensers to convert steam to water to be returned to the boiler for reheating and reusing in the steam generation process. The emission control process reduces GHG emissions (e.g., CO_2, sulfur dioxide, nitrogen oxides, and particulate matter), especially from the coal combustion processes (Figure 2.8). Various emission control technologies, such as scrubbers,

Figure 2.8 Example of emissions released from a coal power plant. Source: Wiley Digital Library.

electrostatic precipitators, and selective catalytic reduction systems, can remove pollutants from the flue gas before releasing them into the atmosphere. Other solutions can include mixing chemicals and additives with conventional coal to reduce the pollution when it is burned.

2.2.1 Carbon Capture, Utilization, and Sequestration (CCUS)

Coal-fired power generation has some advantages, including its abundance and reliability as a fuel source. However, it is also associated with several environmental challenges, including air pollution, GHG emissions, and the release of toxic materials, such as mercury and arsenic. To address these challenges, CCUS strategies can be implemented to capture, utilize, and store GHG emissions (Figure 2.9). CCUS strategies can be applied in a wide range of power generation settings and sources, such as coal, natural gas, and biomass feedstocks. Despite the rapid growth and maturity of renewable energy generation, the world still relies on coal, crude oil, and natural gas to meet the global energy demand. The CCUS technologies are essential during the transition from fossil fuels to renewable energy technologies.

The United States, Russia, and India have the most coal reserves globally. In the United States, coal-fired power plants need to demonstrate their CCUS technologies to get support and tax credits; however, most of the new projects have

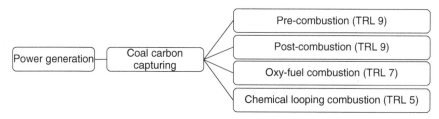

Figure 2.9 Main carbon-capturing technologies from coal-fired power generation plants.

not been successful due to higher costs than other energy sources, such as natural gas. According to the latest U.S. Department of Energy reports, the total cost of solar and wind energy has been reduced by around 85% and 66% in a decade, respectively. Currently, the U.S. Department of Energy, Clean Coal Power Initiative supported 8 coal CCUS projects, and only one of them (Petra Nova project) was completed and entered operation in early 2017, but the project halted operation in 2020. Petra Nova CCUS plant was able to capture 90% of CO_2 emissions from a 240 MW eq. flue gas slipstream in Texas, store up to 1.4 million metric tons of CO_2 annually, transport it to oil fields with 80 miles pipeline, and reuse it for enhanced oil recovery [15]. The other selected CCUS projects withdrew due to less economic viability.

Several carbon-capturing approaches and sorbents to capture carbon and separate CO_2, such as monoethanolamine or modulated amine blend, require 3.5 and 2.1 GJ energy per CO_2 ton, respectively [16]. These approaches can capture carbon before entering the atmosphere and are either stored deep underground or reused for commercial purposes. Recent studies show graphene-type materials have a high CO_2 adsorption capacity (around 0.07 mol/g). Pre/post-combustion technologies present over 96% of existing approaches, and the United States and China are leading most CCUS technologies with 7 and 5 projects out of 22 global projects, respectively [16]. More detailed information can be found at the U.S. Department of Energy, Office of Fossil Energy and Carbon Management.

2.2.1.1 Pre-Combustion/Physical Absorption (TRL 9)

Pre-combustion is one of the main carbon-capturing methods, converting coal into a gas before it is burned, which allows for easier separation of CO_2 from the gas stream. The pre-combustion process involves gasification, water–gas shift (WGS) reaction, and carbon capturing. This process consists of a gasifier to produce synthesis gas (syngas) and a WGS reactor to produce CO_2 and hydrogen from the syngas (Figure 2.10) [16]. The separation process of hydrogen and CO_2 can use chemical or physical absorption techniques through syngas scrubbing.

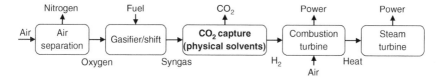

Figure 2.10 Flow diagram of pre-combustion CCUS technology for power generation from coal. Source: Adapted from National Energy Technology Laboratory [17].

Gasification involves converting coal into a gas by reacting it with steam and oxygen in a high-pressure environment. This process produces a gas mixture known as syngas, which consists primarily of hydrogen and CO. WGS reaction converts the CO in the syngas into CO_2 and more hydrogen. This reaction requires the addition of water and a catalyst. After the WGS reaction, the CO_2 capture unit separates CO_2 from the rest of the gas stream using solvent or adsorbent materials. The captured CO_2 can be compressed and transported for storage or utilization. The pre-combustion process for CCUS from coal power plants can capture CO_2 before it is emitted into the atmosphere, making it another effective way to reduce carbon emissions. The pre-combustion process is highly efficient, with up to 90% of CO_2 captured from coal power plants. The main drawbacks of pre-combustion CCUS include: (i) complexity and high cost to implement due to special equipment and infrastructure and (ii) waste handling and management due to additional waste streams, e.g., coal ash and sulfur dioxide. This solution is a mature pathway for CO_2 capturing and has TRL 9, which needs further improvement to stay competitive. The pre-combustion process is much cheaper due to the less energy requirements compared to other CO_2 capturing methods. The main challenge is the overall efficiency that can be improved by different techniques, such as solvent regeneration (or avoiding solvent reduction) using ionic liquids [18]. Despite these challenges, pre-combustion CCUS remains a promising technology for reducing CO_2 emissions from coal power plants.

2.2.1.2 Post-Combustion/Chemical Absorption (TRL 9)

Post-combustion is another process used for carbon capturing from coal power plants, which involves capturing CO_2 from the flue gas produced by burning coal. Post-combustion technology uses a CO_2 absorber unit that contains sorbent (e.g., monoethanolamine solvent) and CO_2 stripper for separating the pure CO_2 from flue gas (Figure 2.11). Post-combustion process for CCUS involves the following four main steps: (i) capturing the flue gas (e.g., carbon, nitrogen, water vapor, and trace amounts of other gases) by burning coal; (ii) cooling the flue gas to reduce its temperature and increase the concentration of CO_2 in the gas stream; (iii) separating the carbon from the rest of the gas stream using solvent or adsorbent materials

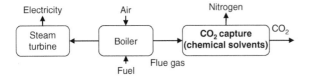

Figure 2.11 Flow diagram of the post-combustion CCUS technology for power generation from coal. Source: Adapted from National Energy Technology Laboratory [17].

(e.g., amine); and (iv) compressing the CO_2 to reduce its volume and make it easier to transport, store, or utilize (e.g., enhanced oil recovery).

Amine is the most common solvent used for post-combustion CCUS, a chemical compound that selectively reacts with CO_2 for separating it from the other gases in the flue gas. The benefits of chemical solvents (amine-based) are high absorption capacity and selectivity at low pressure compared with physical solvents. Recent studies investigated the advantages and disadvantages of different solvents (physical or chemical), absorbents (e.g., calcium carbonate or hydrotalcite), and membranes (e.g., palladium or silica-based) to improve the overall efficiency, CO_2 absorption, and purity [16]. This technology for CO_2 capturing has been used in the chemical industry for almost a century.

Post-combustion is an effective solution for reducing CO_2 emissions from coal power plants using different methods, such as chemical absorption, physical adsorption, and membrane separation. Monoethanolamine chemisorption is the only commercialized method with high capital and operational costs. Membrane separation can enhance sustainability benefits and reduce operational costs. The main challenges of post-combustion CCUS technology for power generation from coal include: (i) high energy requirement for capturing and separating CO_2 that can reduce the efficiency of the power plant; (ii) high cost of the equipment and infrastructure; and (iii) process complexity due to water condensation, selectivity reduction, and membrane temperature adjustment. Post-combustion process is one of the most used methods among other CCUS processes. Earlier studies utilized various solvents (e.g., CO_2BOL, aminosilicones, and amine-organic) to improve thermal stability and efficiency, and lower fuel cost, pressure, energy consumption, and emissions. Several companies and national laboratories have conducted solvent research and development for CCUS, such as Shell, Pacific Northwest National Lab, and General Electric [19].

2.2.1.3 Oxy-Fuel Combustion (TRL 7)

Oxy-fuel combustion is one of the main CCUS processes that involves burning coal in a mixture of pure oxygen and recycled flue gas, which creates a gas stream consisting primarily of CO_2 and water vapor. CO_2 can then be captured and

Figure 2.12 Flow diagram of oxy-fuel combustion CCUS technology for power generation from coal. Source: Adapted from National Energy Technology Laboratory [17].

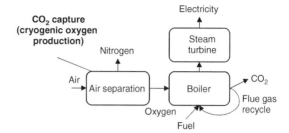

stored or utilized for other purposes. Oxy-fuel combustion technology uses a pure oxygen stream instead of air for combusting carbon-based fuel and separating CO_2. Oxy-fuel combustion process includes an oxygen separator, boiler, particle removal, condenser units, and compressor that can combust the carbon-based fuel using oxygen and send the flue gas to particle removals and condensers to remove water and sulfur and separate CO_2.

The oxy-fuel combustion process for carbon capturing involves several steps (Figure 2.12). The first step is air separation into its component gases, primarily nitrogen and oxygen, using an air separation unit that relies on cryogenic distillation or other separation techniques. The second step is coal combustion in a mixture of oxygen and recycled flue gas. The recycled flue gas is used to provide the necessary volume of gas for combustion and to moderate the temperature of the combustion process. The combustion process generates a gas stream, consisting primarily of CO_2 and water vapor. The third step is CO_2 capturing and separation from the gas stream using a solvent or adsorbent material similar to post-combustion processes.

One of the advantages of oxy-fuel combustion is that CO_2 is already concentrated in the gas stream, making it easier to capture and separate from other gases. Then CO_2 is compressed to high pressure, transported to a storage site, or utilized for other purposes. The oxy-fuel combustion process is straightforward, but the main challenges are the energy required to separate air into its component gases and to compress CO_2. Also, the costs of required equipment and infrastructure are relatively high. The energy-intensive, high-cost process (e.g., cryogenic distillation) for pure oxygen production is the limiting factor, but overall, oxy-fuel technology is an energy-efficient pathway for CO_2 separation and capture. Despite these challenges, oxy-fuel combustion is an effective method for reducing CO_2 emissions from coal power plants. It can achieve high levels of CO_2 capture (up to 90% or more), and the captured CO_2 is already compressed, reducing the energy required for transport and storage. Oxy-fuel combustion can help reduce GHG emissions and mitigate the impacts of climate change. Recent studies investigated several methods for improving air separation, such as ion-transport and oxygen-transport membranes, along with chemical looping that can reduce the required energy and operational costs. Further details about oxy-fuel combustion technology have been provided by Yadav and Mondal [20].

2.2.1.4 Chemical Looping Combustion (TRL 5)

Chemical looping combustion is another CCUS approach for reducing carbon emissions from coal power generation. This solution can capture CO_2 after burning coal without energy-intensive separation processes. The basic science behind chemical looping combustion for CCUS involves using metal oxide particles (e.g., iron or nickel) as the catalyst. In this process, metal oxide (MeO) is circulated between the fluidized reactors as the oxygen carrier to produce CO_2 and water in the reducer (Figure 2.13). Later, methyl (Me) can be separated in the cyclone and reused in other cycles for reducing the process cost.

This process uses two reactors (i.e., air and fuel) for circulating MeO particles. In the air reactor, MeO particles are exposed to air to oxidize and release heat. In the fuel reactor, hot MeO particles react with the coal to generate heat and CO_2 and capture CO_2 by MeO particles. Then loaded MeO particles with CO_2 are transported back to the air reactor to start the cycle again. The captured CO_2 can be removed from MeO particles by heating them to a high temperature, which releases CO_2 and regenerates MeO particles.

The benefits of chemical looping combustion include: (i) high CO_2 capturing level without energy-intensive separation processes and (ii) lower oxygen concentration than traditional combustion processes, which reduces the amount of nitrogen in the flue gas and simplifies CO_2 capture. The challenges include: (i) the

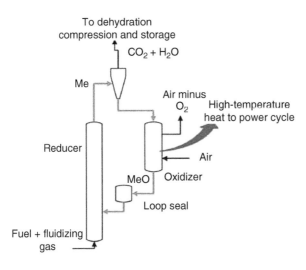

Figure 2.13 Flow diagram of chemical looping CCUS technology for power generation from coal. Source: Bui et al. [21]/Royal Society of Chemistry.

high cost and complexity of the reactor system, which requires two reactors and a circulating fluidized bed and (i) the limited availability of suitable MeO particles, which can be expensive and difficult to produce. Earlier techno-economic studies reported that chemical looping combustion is more cost-effective than oxy-fuel combustion [22]. The key parameters are the lifetime and cost of the oxygen carrier materials. Recent studies show that the latest development reached TRL 5 and can generate power to a scale of around a megawatt [21].

Overall, CCUS technologies have the potential to reduce carbon emissions and mitigate the impacts of climate change. However, each technology has its own set of benefits and drawbacks in terms of sustainability (Table 2.3). The choice of technology will depend on various parameters and variables, such as the specific power plant, coal type, and local environmental regulations. It is essential to continue research and development to improve the sustainability of CCUS technologies and ensure their long-term viability as a tool for reducing carbon emissions.

2.2.1.5 Utilization Strategies (TRL 9)

Globally, anthropogenic CO_2 emission is over 32,000 million tons per year, and the gross CO_2 consumption is around 200 million tons per year [16], which is negligible. Utilization is one of the main components of CCUS strategies, which involves the conversion of captured CO_2 into valuable products, rather than simply storing it underground. The main utilization routes are enhanced oil recovery (Figure 2.14), gas recovery, power generation, district cooling, mineralization, beverage carbonation, food packaging, and chemical/fuel industries. For example, coal plays a vital role in power generation in some countries (e.g., India) with high coal reserves (150 gigatons). They can employ CCUS strategies in coal power plants and integrate them with other thermochemical processes (e.g., gasification). Increasing carbon capturing affects the size and complexity of the WGS process, which can reduce energy efficiency and increase power production costs.

The basic science behind utilization processes for CCUS from coal power plants involves understanding the chemistry and thermodynamics of the conversion reactions. Currently, the largest end-use for carbon capturing is in enhanced oil recovery (EOR) [19]. Enhanced oil recovery is a process that involves injecting CO_2 into oil reservoirs to increase oil production. The injected CO_2 dissolves into the oil through a process called solubility trapping, reducing its viscosity, and making it easier to extract. The solubility of CO_2 in oil is influenced by pressure, temperature, and the composition of the oil. It can be an attractive utilization option for CCUS from coal power plants because it can generate revenue from the sale of the additional oil produced.

Table 2.3 Comparison of leading CCUS approaches from coal-fired power plants.

Approach	Mechanism	Benefits	Challenges	TRL
Pre-combustion	Carbon is removed before combustion: Coal is gasified, and the resulting gas stream is cleaned and separated into CO_2 and hydrogen	High efficiency and yield due to high pressure. Excellent suitability for gasification. It can achieve high levels of CO_2 capture, up to 90% or more. The captured CO_2 is already compressed, making it easier to transport and store	It requires significant modifications to the power plant, like adding a gasification unit. Negative impacts of CO-shift process on heating value and turbine design. Pressurization requirements for post-carbon capturing. Gasification can produce other air pollutants (e.g., SO_2 and NO_x). The energy required for gasification and hydrogen production may offset some of the carbon savings from captured CO_2	Commercialization and deployment stage (8–9)
Post-combustion	Carbon is removed after combustion: Flue gas is cleaned, and CO_2 is captured using a solvent or adsorbent material	It can be used in any existing plant for carbon capturing. Good suitability for gasification. It can be retrofitted to existing power plants with minimal modifications, making it a flexible option. The solvent or adsorbent material used in the capture process can be recycled for reducing waste	Complex and energy-intensive separation process due to low pressure and carbon concentration. CO_2 capture rate is typically lower than that of pre-combustion or oxy-fuel combustion. The solvent or adsorbent materials used in the capture process may have environmental impacts. The captured CO_2 may require additional compression before transportation and storage	Commercialization and deployment stage (8–9)

Oxy-fuel combustion	Fuel is combusted with pure oxygen: Coal is burned in a mixture of oxygen and recycled flue gas, resulting in a gas stream consisting primarily of CO_2 and water vapor, which is then captured	It can achieve high levels of CO_2 capture, up to 90% or more. The captured CO_2 is already compressed, making it easier to transport and store	It requires modifications to the power plant to add the air separation unit and combustion unit. Oxygen and O_2 purification requirements and their associated costs. Poor suitability for gasification. The energy required for the air separation and combustion processes can increase the overall energy consumption	Commercialization stage (7–8)
Chemical looping combustion	Similar to pre-combustion, it involves gasification to produce syngas that is used for combustion and capturing CO_2 from the flue gas after combustion	It can perform both gasification and steam methane reforming in a single unit, unlike pre-combustion that requires the use of separate ones. It can potentially capture other pollutants (e.g., SO_2 and NO_x) along with CO_2 and achieve higher CO_2 capture rates with lower energy requirements. It involves combustion with a limited amount of oxygen to produce a CO_2-rich flue gas	High cost and complexity of the reactors. The limited availability of suitable metal oxide particles can be expensive and difficult to produce	Early stages of development and commercialization (4–5)

Source: Adapted from NETL [19].

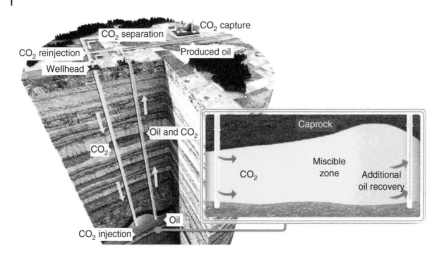

Figure 2.14 Schematic of CO_2 utilization and storage for the enhanced oil recovery. Source: Global Carbon Capture and Storage (CCS) Institute [23].

Other applications include chemical or fuel production, beverage carbonation, mineral carbonization, metallurgy extraction, and machinery [16]. Production of chemicals and fuels is another potential utilization option for CCUS from coal power plants. Captured CO_2 can be converted into chemicals (e.g., methanol) or fuels (e.g., synthetic diesel) through a carbon capture and utilization process that can be an attractive option to generate revenue from byproducts. The captured CO_2 is converted into chemicals or fuels through various processes, such as catalytic conversion, electrochemical conversion, or biological conversion. The specific process used will depend on the desired product and the characteristics of the feedstock.

Mineralization is a process that involves the reaction of captured CO_2 with naturally occurring minerals (e.g., magnesium and calcium silicates) to form stable carbonates. The captured CO_2 reacts with minerals in the presence of water to form carbonates. The reaction kinetics and product formation are influenced by mineral composition, reaction temperature and pressure, and the presence of catalysts. Mineralization can be an attractive utilization option for CCUS from coal power plants because it permanently removes CO_2 from the atmosphere and can potentially generate revenue from the sale of the resulting carbonates. Overall, utilization processes for CCUS from coal power plants require a fundamental understanding of the underlying chemistry and thermodynamics of the conversion reactions. By leveraging this understanding, scientists and engineers can develop efficient and economically viable utilization processes to help reduce carbon emissions.

2.3   Natural Gas

Natural gas is a fossil fuel primarily composed of methane and other hydrocarbons (e.g., ethane, propane, and butane). According to the U.S. EIA, natural gas is the primary source of power generation, with over 38% (around 1,600 kWh) in 2022. Power generation from natural gas involves the combustion of natural gas to produce high-pressure steam for running steam turbines and power generators. The combustion process mixes natural gas with air in the combustion chamber and burns the gas to generate heat and other gases. Then the generated heat can be used to produce high-pressure steam from water, run the steam turbine, and drive the generator to produce power. The other gases from the combustion process are usually released into the air. The advantages of power generation from natural gas are: (i) a relatively clean-burning fossil fuel, emitting less pollutants (e.g., sulfur dioxide, nitrogen oxides, and particulate matter) compared to coal or oil, and (ii) a higher thermal efficiency compared to other fossil fuels for power generation, meaning it can generate more power using the same amount of fuel. The major disadvantage is that the combustion of natural gas still produces CO_2 and other GHGs that contribute to climate change. To mitigate the environmental impacts of natural gas power generation, CCUS technologies can be used similar to coal power plants to capture and sequester carbon emissions (Figure 2.15). There are several approaches that have been investigated, and the main ones are post-combustion/chemical absorption and supercritical CO_2 cycle.

Figure 2.15   Example of advanced post-combustion carbon capture technology from a natural gas power plant with up to 12 MWe and 200 tons CO_2/day. Source: U.S. Department of Energy, National Energy Technology Laboratory (public domain) [24].

2.3.1 CCUS

The basic science of CCUS from natural gas power plants is very similar to CCUS from coal power plants and involves understanding the thermodynamics and kinetics of CCUS processes. But there are some differences for natural gas power CCUS processes due to them being less carbon-intensive compared to other fossil fuel sources (e.g., coal). Particularly, the flue gases from natural gas power plants contain around 4–5% CO_2, but coal-fired flue gas has around 12–15% CO_2 by volume [24]. The main CCUS processes at natural gas power plants involve capture, transport, and storage. The first step is to capture carbon emissions, and there are several different capture technologies available, including post-combustion capture, pre-combustion capture, and oxy-fuel combustion. Post-combustion capture is the most common technology used for natural gas power plants, as it can be retrofitted onto existing plants. This technology involves capturing CO_2 emissions from the flue gas using solvents or solid sorbents. The main carbon capturing parameters are the type of capture technology used, the temperature and pressure of the flue gas, and the type of solvent or sorbent used. The second step is transport to other locations for utilization or storage, and it can be done using pipelines or other transportation methods (e.g., trucks or ships). The main transportation parameters are the distance between the capture site and the storage site, the type of transportation method used, and the safety and security of the transportation process. The last step is to store CO_2 underground in geological formations, such as depleted oil and gas reservoirs, or saline formations, by injecting CO_2 into the formations, and is trapped by the surrounding rock and geological features. The stored CO_2 is monitored over time to ensure that it remains secure and does not leak back into the air. The main storage parameters are geological formation types used for storage, CO_2 injection rate, and the long-term stability of stored CO_2.

2.3.1.1 Post-Combustion/Chemical Absorption (TRL 8)

As explained earlier, post-combustion/chemical absorption is a common method for carbon capturing from coal and natural gas power plants. This process involves capturing CO_2 emissions from the flue gas using a solvent, such as an aqueous amine solution (Figure 2.16). The basic science behind these processes involves chemical and physical interactions among the flue gas, the solvent, and CO_2. The main two steps include: (i) the flue gas from the natural gas power plant goes through an absorber tower for spraying the solvent into the gas stream and (ii) the solvent reacts with CO_2 in the gas, forming a chemical bond and removing CO_2 from the gas stream. For example, the amine solvent acts as a weak base, reacting with acidic CO_2 to form a chemical bond. The chemical reaction is reversible,

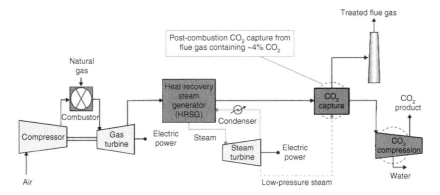

Figure 2.16 Schematic of post-combustion/chemical absorption processes for CCUS from natural gas power plant. Source: U.S. Department of Energy, National Energy Technology Laboratory (public domain) [24].

so the solvent can be regenerated by heating it and releasing CO_2, which is then stored or used for other purposes.

The key physical interaction parameter in the post-combustion process is the CO_2 mass transfer rate from the flue gas into the solvent. The mass transfer rate depends on the CO_2 concentration in the flue gas, gas temperature and pressure, and solvent properties such as viscosity and surface tension. The efficiency of the post-combustion/chemical absorption process depends on gas concentration and flow rate, solvent type and flow rate, and absorber tower operating conditions (e.g., temperature and pressure). The National Energy Technology Laboratory has provided detailed information about the post-combustion/chemical absorption process for CCUS from natural gas power plants [24].

2.3.1.2 Supercritical CO_2 Cycle (TRL 6)

Supercritical CO_2 cycle is a novel technology that can be used for CCUS from natural gas power plants. The basic science of this technology involves the thermodynamics of a fluid in a supercritical state and the operation of a closed-loop cycle. Supercritical CO_2-based power cycles operate similarly to other steam-based power cycles; however, they use CO_2 as the working fluid. During power generation from natural gas, the exhaust gas from the turbine is directed through a heat exchanger, where it transfers its heat to a supercritical CO_2 fluid, which is maintained at high pressure and temperature (above its critical point). This results in unique thermodynamic properties, such as high density and low viscosity, which make it an excellent working fluid for power generation. The CO_2 fluid absorbs the heat from the exhaust gas and expands through a turbine to generate additional power. The expanded CO_2 fluid is then cooled and compressed back to its initial state,

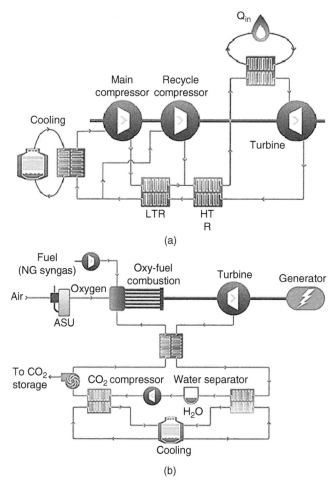

Figure 2.17 Schematic of supercritical CO_2-based power cycles: (a) indirectly heated closed-loop Brayton cycle and (b) directly heated cycle. Source: U.S. Department of Energy, National Energy Technology Laboratory (public domain) [25].

ready to begin the cycle again. The captured CO_2 from the flue gas can be injected into geological formations or used in industrial processes. Figure 2.17 shows both directly and indirectly heated supercritical CO_2-based power cycles.

This technology has shown potential due to its high heat-to-power process efficiency, power density, and similarity to the existing turbine cycles [26]. The process efficiency can be improved by optimizing the operating conditions of the system, such as the temperature and pressure of supercritical CO_2 fluid, fluid flow rate, and the type of heat exchanger used. Other benefits are lower capital cost, higher thermal efficiencies, and reduced water and fuel use than traditional steam cycles [27]. Also, a small footprint compared to other CCUS technologies makes this technology suitable for retrofitting onto existing natural gas power plants. This technology can be implemented in directly or indirectly heated applications. The direct method mixes fuels (e.g., natural gas or coal) with CO_2 as the working fluid to run the turbine and generate power. The indirect method uses or recovers heat sources from the boiler to indirectly heat CO_2 through a heat exchanger and run the turbine or preheat the compressed CO_2. More detailed information about the supercritical CO_2 cycle process has been provided by the National Energy Technology Laboratory [25].

Overall, each CCUS process for capturing carbon emissions has benefits and drawbacks. The main applied carbon-capturing methods from natural gas power plants are post-combustion/chemical absorption, supercritical CO_2 cycle, and oxy-fuel combustion. Post-combustion/chemical absorption process is the most common CCUS method due to its simplicity, low energy consumption, and compatibility with existing power plants, but the drawbacks are its high capital cost, solvent requirements, and wastewater generation. The benefits of the supercritical CO_2 cycle process include its high efficiency, low water consumption, and small footprint. The main drawback is the need for advanced materials for a supercritical (high pressures and temperatures) process. The benefits of the oxy-fuel combustion process include its high purity of captured CO_2, compatibility with existing power plants, and relatively low capital cost. The drawbacks are oxygen requirement and corrosive gas stream.

2.4 Biomass

Biomass-based energy (bioenergy) is a renewable solution that can replace transportation fuel and power generation from crude oil, natural gas, and coal. Growing carbon-rich biomass produces near-zero emissions through carbon circulation from the atmosphere to plants as long as consumption and replenishment stay in balance. The basic science of power generation from biomass feedstocks involves converting the energy stored in biomass into power using various processes, either thermochemical or chemical. Biomass feedstocks consist of organic materials from wood chips, crop residues, agricultural waste, and municipal solid waste.

Biomass power generation involves biomass preparation, conversion process, steam generation, and cooling. In particular, biomass feedstocks are first collected, sorted, and processed to remove contaminants, such as rocks, soil, and noncombustible materials. The feedstocks are then shredded, chipped, or ground into a uniform size for efficient combustion. The pretreated biomass feedstocks, after dewatering and size reduction, are burned in a boiler to produce high-pressure steam. The thermochemical (combustion) conversion process can be either direct combustion, pyrolysis, or gasification, depending on the type of biomass and the desired energy output. The steam generated from the thermochemical process is directed to a steam turbine, which is connected to a generator. The steam turbine converts the thermal energy of steam into mechanical energy, which drives the generator to produce power. The steam that passes through the steam turbine is then cooled using a condenser (cooling tower) that can convert the steam back into water. The water is then returned to the boiler to be reheated and reused in the steam generation process.

There are several drawbacks to the first generation of bioenergy from food crops (e.g., corn and sorghum) due to food and water security, as well as energy requirements for production processes. To manage these drawbacks and reach a net-zero energy future, we need to shift practices and new regulations. For example, recent studies reported that perennial plants (woody and herbaceous) with short lifetimes (around 2 or 3 years) are better resources for bioenergy production due to their high productivity and low water consumption in comparison to food crops. The prior techno-economic studies show that power production from perennial plants can reach $0.014 per kWh by 2050. The average cost for power generation from fossil fuels is around $0.049 per kWh. Biomass power generation has several advantages, including its renewable resources, low carbon emissions, and potential to reduce waste. However, it also has some challenges, including variability in feedstock quality and availability, high initial capital costs, and potential land-use conflicts. To overcome these challenges, biomass power generation can be combined with other renewable energy sources, such as solar, wind, and hydropower, to provide a more stable and reliable power supply. Additionally, bioenergy production (e.g., power, heat, or biofuels) with CCUS processes can capture and sequestrate any CO_2 released when using the bioenergy and any remaining biomass carbon that is not in the liquid fuels.

2.4.1 CCUS

The science of CCUS processes in biomass power generation involves capturing CO_2 emitted during the combustion of biomass feedstocks and either storing it underground or utilizing it for other purposes (Figure 2.18). As explained earlier for power generation from coal or natural gas, CCUS strategies include three main steps: carbon capturing, utilization, and sequestration/storage. The mature CCUS technologies for biomass power generation are post-combustion and pre-combustion. Pre-combustion capture involves gasifying biomass feedstocks before combustion and capturing the CO_2 during the gasification process.

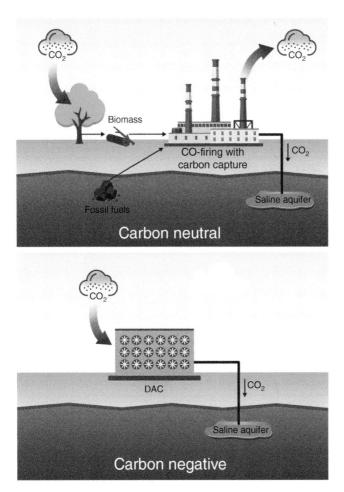

Figure 2.18 Schematic of biomass use for CCUS and power generation. Source: U.S. Department of Energy, National Energy Technology Laboratory (public domain) [28].

Post-combustion capture involves separating CO_2 from the flue gas using chemical solvents or physical adsorbents. Additionally, oxy-fuel combustion has potential and consists of burning the biomass in an oxygen-rich environment, which produces a flue gas with a higher concentration of CO_2 that is easier to capture. Biomass power generation with CCUS has several potential benefits, such as low carbon footprint, job creation, and revenue streams from renewable energy and carbon utilization. However, the main drawbacks are high capital costs and deployment barriers.

2.4.1.1 Post-Combustion/Chemical Absorption (TRL 8)

Post-combustion/chemical absorption process for CCUS from biomass power generation is one of the most widely used technologies for capturing CO_2 emissions. The basic science of this process involves using a solvent to capture CO_2 from the flue gas generated during the combustion of biomass feedstocks. The captured CO_2 is then separated from the solvent and either stored or utilized for other purposes.

The post-combustion/chemical absorption process involves: (i) flue gas treatment for removing particulate matter, sulfur dioxide, and other impurities from biomass combustion and reducing viscosity and corrosion; (ii) CO_2 absorption, using an absorption tower and solvents (e.g., monoethanolamine or ammonia) that can react with CO_2 to form a chemical compound and separate it from the flue gas; (iii) CO_2 separation from the CO_2-rich solvent using the stripping column, while the solvent is recycled back to the absorption tower; and (iv) CO_2 compression to a supercritical state and transport it to storage sites (e.g., depleted oil and gas reservoirs or saline aquifers).

The post-combustion/chemical absorption process is a mature technology (TRL 8) that has been used in the industry for several decades. It is relatively simple to operate and can achieve high CO_2 capture rates of up to 90%. However, it has some drawbacks, such as high energy, water, and solvent requirements, as well as the potential for solvent degradation and corrosion.

2.4.1.2 Pre-Combustion/Physical Absorption (TRL 3)

Pre-combustion/physical absorption process for CCUS from biomass power generation is a promising technology for capturing CO_2 emissions. The basic science of this process involves converting biomass feedstocks into syngas, which are treated to remove impurities and capture CO_2. Then CO_2 is separated from the syngas and stored or utilized for other purposes.

The pre-combustion/physical absorption process involves the following steps:

- Gasification: The biomass feedstock is converted into syngas through gasification, which involves heating the biomass in the presence of a gasifying agent (e.g., oxygen, steam, or air). The syngas consists of CO, hydrogen, and other impurities.
- Shift conversion: The syngas is then subjected to a shift conversion process, where CO is reacted with steam to produce more H_2 and CO_2. This step increases the concentration of CO_2 in the syngas and prepares it for CO_2 capture.
- CO_2 capture: CO_2-rich syngas is directed to a physical absorption column, where it is treated with a solvent (e.g., Selexol or Rectisol) to capture CO_2. The solvent reacts with CO_2 to form a chemical compound that can be separated from the syngas.

- Separation: CO_2-rich solvent is directed to a stripping column, where heat is applied to separate CO_2 from the solvent. Then CO_2 is compressed and transported for storage or utilization, while the solvent is recycled back to the absorption column.
- Power generation: The clean syngas is then used to fuel a gas turbine or internal combustion engine to generate power. The generated power can be used on-site or fed into the grid.

The benefits of pre-combustion/physical absorption process include: (i) high CO_2 capture rate (up to 95%), (ii) low energy requirements, (iii) higher efficiency due to the use of byproducts (e.g., syngas) for other applications (e.g., heating or transportation fuel), and (iv) the potential for co-production of hydrogen. The drawbacks include: (i) high water and solvent needs, (ii) high potential for solvent degradation and corrosion, and (iii) complex equipment and a higher capital investment, which may limit its scalability.

Compared to pre-combustion/physical absorption, the main disadvantages of post-combustion/chemical absorption include: (i) higher energy requirements, (ii) higher operational costs, and (iii) higher environmental impacts due to chemical solvents used for capturing CO_2 and generated wastes. Compared to other methods, the oxy-fuel combustion process is more sustainable than other processes because it has high efficiency, produces a flue gas stream with a high CO_2 concentration, and is easier to capture. The main disadvantages are high oxygen requirement and high capital costs, which may limit its scalability. Overall, the sustainability benefits of CCUS processes for biomass power generation depend on various factors, such as biomass feedstock type, plant size, location, and economic considerations.

2.5 Ammonia

Globally, the primary energy sources in the industrial sector are fossil fuels (e.g., coal, natural gas, and petroleum), which require crucial steps toward decarbonization. Ammonia (NH_3) is a colorless gas and carbon-free energy carrier that can be used for heat and power generation using various technologies, such as combustion engines, gas turbines, fuel cells, or co-firing with fossil fuels (Figure 2.19) [29]. The byproducts of ammonia combustion or co-firing with fossil fuels (e.g., coal and natural gas) are nitrogen and water, which are environmentally friendly and can reduce carbon emissions. Ammonia can also increase process efficiency and be used as feedstock to generate syngas by reacting ammonia with air or oxygen and forming nitrogen, hydrogen, and CO, that can run gas turbines for power generation. Fuel cells can break down ammonia into nitrogen and hydrogen ions and

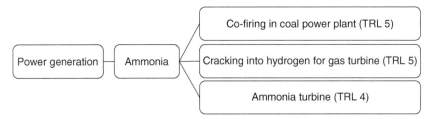

Figure 2.19 Main technologies for power generation from ammonia.

react with oxygen to produce water and power. Ammonia fuel cells have a high energy density, and their byproduct is water.

Ammonia can be used as an energy storage medium in the form of ammonia borane, which is a solid material that releases hydrogen when heated. Ammonia borane (BH_6N) can be used to store hydrogen for fuel cells or for power generation using a combustion engine. Recent studies have investigated the feasibility of ammonia as a potential energy carrier for decarbonization and power generation [30, 31]. Besides, chemical energy storage cost, using ammonia is 26–30 times cheaper than hydrogen due to several constraints on the hydrogen economy, such as handling, infrastructure, and high flammability range [31]. Ammonia can be stored under easy conditions, e.g., atmospheric temperature with 116–145 psi or −33 °C with atmospheric pressure (14–15 psi) [31]. Globally, the ammonia market is around 180 million tons per year as the second most commercialized chemical due to high applications (Figure 2.20).

The Haber–Bosch is the primary process for ammonia synthesis. Once ammonia is produced, various methods (e.g., combustion) can be used for power and heat generation. Ammonia has the potential as fuel in transportation because it requires less storage volume than hydrogen (2.5 times higher liquid volume density), and GHG emissions from the ammonia-driven vehicle are lower than gasoline and diesel [31]. The main barrier for ammonia to be a global energy source is public acceptance due to its bad odor. Co-firing dual fuels (e.g., hydrogen and ammonia) can be a good approach to address the existing decarbonization challenges, such as emission mitigation. Hydrogen can be an ideal fuel to improve the ammonia flammability range and combustion applications. Other fuels (e.g., ethanol and kerosene) can be suitable promoters for ammonia combustion; however, the main issues are low combustion temperatures and high emissions [31].

The main benefits of power generation from ammonia are: (i) ammonia can be produced from renewable energy sources, such as wind or solar power, making it a renewable energy carrier; (ii) ammonia produces nitrogen and water vapor as byproducts, which are environmentally friendly and do not contribute to GHG emissions; (iii) ammonia has a high energy density, which means it can store more

Figure 2.20 Example of a chemical plant for ammonia production. Source: Wiley Digital Library.

energy per unit volume than other fuels, such as hydrogen or methane; (iv) the existing infrastructure can be used for ammonia production and transport, which can be leveraged for power generation; and (v) ammonia can be used for power generation in a variety of ways, including combustion, fuel cells, and synthesis gas production. The major drawbacks of power generation from ammonia are: (i) ammonia is toxic and can be dangerous if not handled properly, which can be a challenge for large-scale production and transportation; (ii) ammonia is highly volatile and requires specialized storage and transportation infrastructure to prevent leaks and accidents; (iii) ammonia has high ignition energy, which can make it challenging to use as a fuel in specific combustion engines and gas turbines; (iv) the production and transportation of ammonia can be costly, and the cost of ammonia-based power generation technologies may be higher than conventional fossil fuel-based power generation; and (v) many of the ammonia-based power generation technologies, such as ammonia fuel cells, are still in the development stage, which may delay their commercialization and widespread adoption. Overall, ammonia has the potential to be an essential energy carrier for power generation. However, more research and development are needed to improve the efficiency and reduce the costs of ammonia-based power generation technologies. Further information about power generation from ammonia has been provided by [29].

2.5.1 Co-firing in Coal Power Plant (TRL 5)

Co-firing ammonia in coal power plants involves mixing ammonia with coal and burning the mixture to generate power. The main steps and science behind this process involve combustion, emissions reduction, and fuel blending. Since coal and ammonia are combustible materials, mixing them forms a fuel blend that can be used for power generation in the combustion chamber. The high hydrogen content of ammonia makes it a good fuel for combustion. During combustion, ammonia reacts with oxygen and forms nitrogen and water vapor, which are carbon-free byproducts and help coal power plants reduce their GHG emissions and address their existing environmental challenges, especially high GHG emissions released into the air. The byproducts (nitrogen and water) can be used in different processes, such as coal combustion or WGS reaction. For example, nitrogen can be used as an inert gas to dilute oxygen amount for coal combustion, leading to reduce CO_2 emissions. The main factors in this process are coal type, combustion conditions, and emissions regulations. The optimum blending ratio needs to be determined experimentally to ensure stable combustion and minimum emissions. Additionally, co-firing ammonia in coal power plants may require modifications to the plant's infrastructure, such as the combustion chamber and fuel handling systems, to accommodate the different fuel properties of ammonia. Also, safety measures must be in place to safely handle ammonia during storage, transportation, and blending.

Using ammonia as a fuel for co-firing has several problems due to its main characteristics (e.g., high ignition energy, low flame temperature, and low combustion speed) compared to other fuels. Several studies investigated ammonia co-firing in coal-fired thermal power plants. They concluded that substituting coal with ammonia can reduce the concentration of coal particles and CO_2 emissions, affecting flue gas composition and combustion performance, which also depends on the process configuration (e.g., furnace) and mixing ratios (co-firing strategy). Ammonia characteristics are not a good fit for combustion, and the main drawbacks are efficiency and high carbon in fly ashes. Overall, co-firing ammonia in coal power plants can be an effective way to reduce GHGs and increase the efficiency of the combustion process. However, further investigation is needed to optimize the blending ratio, ensure safe ammonia handling, and reduce the cost of ammonia production and transport. More detailed information about co-firing ammonia in coal power plants has been provided by Valera-Medina et al. [31].

2.5.2 Cracking into Hydrogen for Gas Turbine (TRL 5)

Cracking ammonia into hydrogen for gas turbines involves steam methane reforming to break down ammonia into its constituent elements (i.e., nitrogen and hydrogen). The basic science of cracking ammonia into hydrogen for gas turbines includes steam methane reforming, ammonia cracking, gas turbine combustion, and emission reduction. Steam methane reforming is a chemical process that involves reacting steam and methane with a catalyst to produce hydrogen, CO, and CO_2. In the case of cracking ammonia, the process is modified to replace methane with ammonia. Ammonia cracking involves passing ammonia over a catalyst at high temperatures, usually around 800–900 °C, to break down the ammonia into nitrogen and hydrogen. The reaction can be represented as: $2NH_3 \rightarrow N_2 + 3H_2$. Burning the generated hydrogen from this process produces steam and heat that can run gas turbines for power generation. Cracking ammonia into hydrogen for gas turbines can help reduce GHG emissions compared to conventional fossil fuel-based power generation. Hydrogen combustion produces only water vapor as a byproduct, which does not contribute to GHG emissions.

Cracking ammonia into hydrogen for gas turbines may require modifications to the gas turbine's combustion chamber and fuel handling systems to accommodate the different fuel properties of hydrogen. Additionally, safety measures must be in place to safely handle ammonia during storage, transportation, and cracking. Gas turbines have been used for energy production for decades, using three main components (i.e., compressor, combustor, and turbines) and depending on process pressure and temperature. Ammonia can be used as a hydrogen carrier and stored for long periods without degradation. Therefore, several studies investigated ammonia cracking into hydrogen for power generation through running gas turbines. Their results show that ammonia requires higher ignition energy than other fuels due to several shortcomings, such as slow kinetics and narrow stability range. Currently, the main fuel cells that can generate power from ammonia are solid-oxide, proton-conducting, and alkaline fuel cells [32]. Overall, cracking ammonia into hydrogen for gas turbines can be an effective way to reduce GHGs and increase the efficiency of power generation. However, more studies are needed to optimize the cracking process, ensure the safe handling of ammonia, and reduce the cost of ammonia production and transport. Further information about cracking ammonia into hydrogen for gas turbines has been provided by [33, 34].

2.5.3 Ammonia Turbine (TRL 4)

Ammonia gas turbines use ammonia as a working fluid for power generation, similar to traditional gas turbines. The main components of this process include the air compressor, combustion chamber, turbine, and generator. The air compression process increases air pressure and temperature for mixing with ammonia and burning in the combustion chamber. The released heat from these processes can run the ammonia turbine and power generator. The process efficiency mainly depends on the turbine design, combustion chamber, mixed

Table 2.4 Comparison of mature technologies for power generation from ammonia.

Technology	Operating Principles	Benefits	Drawbacks
Co-firing in coal power plants	• Mixing ammonia with coal and using the mixture as a fuel source in coal-fired power plants • Reducing emissions of nitrogen oxides by reacting with them to form nitrogen and water vapor	• Lower cost compared to other methods • Used in existing coal power plants	• Not a clean-burning fuel • Limited reduction in emissions compared to other methods
Cracking into hydrogen for gas turbine	• Converting ammonia into hydrogen and using it as a fuel source for gas turbines by passing ammonia through a catalyst to break it down into nitrogen and hydrogen, which can then be used as fuels for gas turbines	• Clean-burning fuel (produces only water vapor as a byproduct) • Higher efficiency compared to ammonia turbines	• Requires additional equipment for ammonia cracking • Higher cost compared to ammonia turbines
Ammonia turbines	• Using ammonia as a working fluid to generate power • Consisting of a compressor, a combustion chamber, a turbine, and a generator	• Clean-burning fuel (produces only nitrogen and water vapor as byproducts) and can be produced from renewable sources • Suitable for power generation in regions where air pollution is a concern	• Lower efficiency compared to traditional gas turbines

fuel quality, and operating conditions. The main benefits of using ammonia as a working fluid in gas turbines include reducing carbon emissions, producing byproducts (e.g., nitrogen and water vapor) that can be used in other processes, and addressing environmental concerns. Also, ammonia can be produced from renewable sources, such as wind and solar power, which can generate power without relying on fossil fuels and reduce GHG emissions. Ammonia-based turbines or combustion engines are not ready to be commercialized due to their physical characteristics and safety issues (e.g., corrosion and toxicity) compared to turbines or engines that use conventional fuels. Overall, the basic science of ammonia turbines is similar to that of traditional gas turbines. Still, further investigation is necessary on ammonia injection ratios to improve the efficiency and operational cost on the commercialization scale. More detailed information about ammonia gas turbines has been provided by [35].

In summary, ammonia turbines, cracking into hydrogen for gas turbines, and co-firing in coal power plants are mature pathways using ammonia as a fuel source for power generation, but they differ in their basic operating principles and efficiency (Table 2.4).

2.6 Case Study: CCUS-Hybrid Energy Systems

According to the Intergovernmental Panel on Climate Change (IPCC), an average carbon capturing rate of 10 GT/year by 2050 is required to limit mean global temperature rise to 2 °C [21, 36]. Therefore, nature-based CO_2 removal and CCUS technologies are crucial in facilitating decarbonization and meeting net-zero targets (Figure 2.21). CCUS hybrid energy systems have become increasingly important

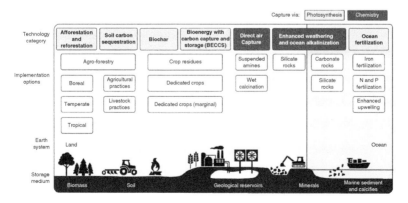

Figure 2.21 Key carbon removal technologies. Source: Minx et al. [37].

in addressing climate change concerns. As the world's energy demands continue to rise, fossil fuels (e.g., natural gas and coal) are still the primary source of power generation. As a result, it is essential to find technologies and pathways to reduce the carbon emissions associated with fossil fuels. CCUS hybrid technologies for power generation have emerged as a promising solution that can help us move towards a low-carbon future. These technologies combine different CCUS technologies to achieve maximum efficiency while reducing carbon emissions.

Some mature carbon-capturing processes for utilization and storage are as follows:

- Pre-combustion capture: This process involves capturing CO_2 from the fuel before it is burned. The fossil fuel is converted into a gas, and the CO_2 is separated from other gases before combustion.
- Post-combustion capture: This process involves capturing CO_2 from the flue gas after the combustion process in the power plant. The captured CO_2 is then separated from other gases and stored underground.
- Oxy-fuel combustion: This process involves burning fossil fuels with pure oxygen instead of air. The resulting flue gas is almost pure CO_2, which makes it easier to capture and store underground.
- Chemical looping combustion: This process involves using MeO to capture CO_2 emissions by circulating MeO between fuel and air streams to capture CO_2 and release it in a concentrated form for storage underground.

These carbon-capturing technologies can be integrated with other processes (e.g., gasification and pyrolysis) for power, heat, or fuel generation. For instance, the integrated gasification combined cycle (IGCC) with carbon capture is an example of a CCUS hybrid energy system, in which coal or other solid fuels are converted into a gas and can be used for power generation. This process includes a couple of steps: carbon capturing, compressing, transporting (via pipeline or other methods), utilization, and storing (underground geologic formation). In addition to sequestering CO_2 emissions, IGCC, with carbon capture, utilizes the captured CO_2 for enhanced oil recovery by injecting the CO_2 into oil reservoirs to help extract additional oil. Overall, IGCC with carbon capture is a hybrid technology that reduces GHG emissions from power generation and provides a means for utilizing captured CO_2.

Supercritical pulverized coal with carbon capture (SCPC) is another CCUS hybrid technology for power generation. The basic science of SCPC with carbon capture involves a range of disciplines, such as thermodynamics, fluid mechanics, and materials science. The main steps are burning coal in a boiler at high temperatures and pressures, capturing the resulting CO_2, and storing it in underground geological formations. SCPC can reduce carbon emissions from coal-fired power plants. Particularly, the SCPC power generation pathway begins with pulverizing coal into a fine powder and burning it in a boiler at temperatures and pressures above the critical point of water (around 374 °C and 22 MPa), allowing the water to exist as a supercritical fluid and extract more energy from the coal. Then the generated steam runs the turbine and generates power. The post-combustion capture system can capture released CO_2 in a separate unit from downstream of the boiler. This process uses solvents to separate CO_2 from the flue gas stream. The main efficiency parameters for the post-combustion capture system are coal properties, boiler design, and solvent selection. The critical point of water is crucial and allows water to act as a solvent and heat transfer fluid. SCPC with carbon capture is a promising technology for reducing GHG emissions from coal-fired power plants; however, it needs further improvement to increase their efficiency and reduce costs.

Natural gas combined cycle (NGCC) power plant with carbon capture is another example of CCUS-hybrid technologies. NGCC process involves burning natural gas to generate steam, run the turbine, and generate power. Then the exhaust gases from this process pass through the carbon capture system. The captured carbon can be utilized in various industrial processes, such as chemical or fuel production. NGCC power plants with carbon capture can achieve high CO_2 capture efficiency, up to 90% or more, without significantly increasing operating costs. Table 2.5 compares the avoided CO_2 capturing costs for power generation through pre- and post-combustion technologies [19]. In addition, captured CO_2 can be used for various applications (e.g., enhanced oil recovery or producing chemicals, fertilizers, and carbonated beverages) to offset the associated cost of capturing.

Earlier studies investigated the CO_2 removal potential and cost for different approaches (Table 2.6).

Other CCUS methods are absorption, adsorption, membrane, and cryogenic processes [19]. CCUS technologies can be integrated with power generation and biofuel production processes and deliver low-carbon power, heat, and

Table 2.5 Comparison of IGCC, SCPC, and NGCC for power generation with and without carbon capturing.

Technology	Source	CO_2 capturing method	CO_2 emissions without capturing (kg/MWH)	CO_2 emissions with capturing (kg/MWH)	Capture rate (%)	CO_2 avoided cost ($/ton)
Integrated gasification combined cycle	Coal	Pre-combustion	793	115	85	43
Supercritical pulverized coal	Coal	Post-combustion	804	109	86	59
Natural gas combined cycle	Natural gas	Post-combustion	370	55	85	80

Source: NETL [19].

Table 2.6 Comparison of key nature-based carbon removal methods.

Approach	CO_2 removal potential (Gt CO_2) cumulative by 2050	Cost ($/t CO_2)
Afforestation/reforestation	147–260	20–100
Biochar	477	0–135
Bioenergy with CCUS	459	45–250
Direct air capture	1,000	40–600
Enhanced weathering	367	20–1000
Ocean fertilization	55–1,027	500
Ocean liming	100	72–159
Soil carbon sequestration	104–130	0–100

Source: IPCC [38], Caldecott et al. [39], NRC [40].

transportation. Several mature CCUS technologies were explained for power and hydrogen generation from coal, natural gas, and biomass earlier, such as pre-combustion, post-combustion, oxy-fuel combustion, chemical looping combustion, supercritical CO_2 cycle, and utilization (Figure 2.22).

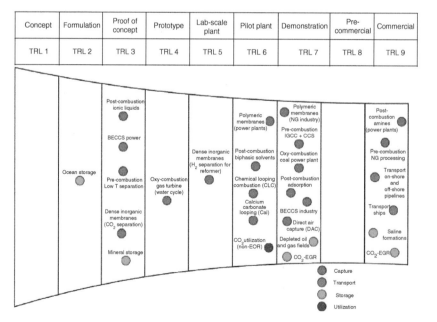

Figure 2.22 CCUS technology development progress. Source: Bui et al. [21].

Table 2.7 Hydrogen production facilities with CCUS.

Company or Project Name	H$_2$ Production (Million ft^3/day)	Technology	CO$_2$ removal rate (%)	Status	Location
Air Products and Chemicals, Inc.	200	SMR	60	Operating	USA
Air Liquide	45	SMR	60	Operating	France
Shell	191	SMR	50	Operating	Canada
H-Vision	636	ATR	88	Under development	Netherlands
HyNet	90	ATR	97	Under development	UK
H$_2$1	2,900	ATR	94	Under development	UK
Acorn Hydrogen	48	ATR	98	Under development	Scotland
British Petroleum	250	–	98	Under development	UK
Air Products Alberta	623	ATR	95	Under development	Canada
Air Products Louisiana	750	ATR	95	Under development	USA

Source: Adapted from Lewis et al. [41].

The key CCUS factors for different applications include CO_2 capture amount, technology, compression, and utilization. For power generation, high CO_2 capturing amount can reduce energy efficiency and increase power generation costs. Selexol and Rectisol processes are the standard technologies for syngas cleanup and CO_2 separation. Selexol can be more efficient in effluent concentrations than Rectisol. For chemical production, CO_2 capture amount affects the size and complexity of WGS. The standard commercial technology is Rectisol that can capture carbon, sulfur, and hydrocarbons (e.g., NH_3 and HCN). For both power and chemical production, CO_2 compression depends on the end-use applications, such as utilization or storage. For example, enhanced oil recovery is the main form of CO_2 utilization and storage, with over 140 active projects globally, mostly in the United States. [21].

Earlier studies reported the CCUS technology development progress. Currently, the United States has the most in-operation CCUS projects. Century Plant and Shute Creek Gas have the highest carbon capture capacity, with 8.4 and 7 $MtCO_2$/year, respectively [21]. Stream methane reforming (SMR) and natural gas autothermal reforming (ATR) are two high TRL methods for hydrogen production using CCUS technologies. Table 2.7 presents the latest operating and under-development hydrogen production facilities with CCUS.

In summary, the benefits of CCUS hybrid systems for power generation are: (i) reduced carbon emissions by capturing and storing CO_2 emissions underground; (ii) increased efficiency of power generation by reducing energy loss associated with CO_2 capture and storage; and (iii) economic benefits by creating new industries and job opportunities. The challenges include: (i) high cost compared to traditional fossil fuel-based power generation technologies, which is mainly due to the high cost of capturing and storing CO_2 emissions; (ii) limited storage capacity since underground storage capacity for large volumes of CO_2 is limited; (iii) energy loss and reduced efficiency of power generation; and (iv) technological barriers to deploy on a large scale. The utilization of captured CO_2 in various applications can generate revenue streams and support the development of new industries. The captured CO_2 can also be utilized in various applications, such as enhanced oil recovery and chemical production, further increasing the overall efficiency of power generation.

Self-Check Questions

1. What is the basic science of nuclear power generation?
2. What are the key parameters to control nuclear energy production?
3. What are the leading nuclear power generation technologies?
4. How does the light-water nuclear reactor work?

5. What are the benefits and challenges of light-water reactors for power generation?
6. How does sodium-cooled fast nuclear reactor work?
7. What are the main environmental impacts of sodium-cooled fast nuclear reactors for power generation?
8. What are the main advantages and disadvantages of sodium-cooled fast nuclear reactors for power generation?
9. How does a high-temperature nuclear reactor work?
10. What are the benefits of high-temperature nuclear reactors for power generation?
11. What are the challenges of high-temperature nuclear reactors for power generation?
12. How does a small modular nuclear reactor work?
13. What are the benefits of small modular nuclear reactors for power generation?
14. What are the challenges of small modular nuclear reactors for power generation?
15. What are the main environmental impacts of small modular nuclear reactors for power generation?
16. What are the differences between fission and fusion processes?
17. What are the benefits of fusion reactors for power generation?
18. What are the drawbacks of fusion reactors for power generation?
19. What are the main steps in power generation from coal?
20. What are the most used, mature carbon-capturing technologies from coal, biomass, and natural gas?
21. What is the pre-combustion/physical absorption process for carbon capturing from coal? And how does it work?
22. What are the benefits and drawbacks of the pre-combustion/physical absorption process for carbon capturing from coal?
23. What is the post-combustion/chemical absorption process for carbon capturing from coal?
24. What are the main steps in the post-combustion/chemical absorption process for carbon capturing from coal?
25. What are the main challenges of post-combustion CCUS technology for power generation from coal?
26. What are the most used post-combustion methods?
27. What is the oxy-fuel combustion process for carbon capturing from coal?
28. What are the main steps of oxy-fuel combustion for carbon capturing from coal?
29. What are the main components of the oxy-fuel combustion process for carbon capturing from coal?

30. What are the benefits and drawbacks of the oxy-fuel combustion process for carbon capturing from coal?
31. What is the chemical looping combustion process? And how does it work?
32. What are the main components of the chemical looping combustion process for carbon capturing from coal?
33. What are the benefits and challenges of the chemical looping combustion process for carbon capturing from coal?
34. What is utilization in CCUS strategies?
35. What are the most used utilization applications?
36. What are the advantages of power generation from natural gas?
37. What is the main disadvantage of power generation from natural gas? And what are the solutions to address it?
38. What are the main carbon capturing, transportation, and storage parameters?
39. What are the main steps of the post-combustion/chemical absorption process for carbon capturing from the natural gas power plant?
40. What are the main parameters in the post-combustion/chemical absorption process for carbon capturing at the natural gas power plant?
41. What is the supercritical CO_2 cycle? And how does it work?
42. What are the main carbon-capturing methods applied to natural gas power plants? And what are their benefits and drawbacks?
43. What are the main steps for power generation from biomass feedstocks? And how does each step work?
44. What are the main steps of the post-combustion/chemical absorption process for carbon capturing from biomass power generation?
45. What are the main steps of the pre-combustion/physical absorption process for carbon capturing from biomass power generation?
46. What are the benefits and drawbacks of the pre-combustion/physical absorption process for carbon capturing from biomass power generation?
47. What are the main technologies for power generation from ammonia?
48. What are the benefits of power generation from ammonia?
49. What are the drawbacks of power generation from ammonia?
50. What are the key carbon removal technologies?
51. What are the examples of CCUS hybrid energy systems for power generation?
52. What are the benefits and challenges of CCUS hybrid energy systems for power generation?

References

1 U.S. EIA (2022). *Annual Energy Outlook*.
2 IAEA (2021). *Nuclear Technology Review*.

3 Forbes (2021). *New Nuclear Fuel Can Be Here Even Faster Than New Reactors.*

4 Forbes (2021). *Nuclear Reactors Could Provide Plentiful Zero-Carbon Hydrogen, If Only We Let Them.*

5 U.S. DOE (2022). *Versatile Test Reactor.* https://www.energy.gov/ne/versatile-test-reactor.

6 U.S. DOE (2022). *Light Water Reactor Sustainability (LWRS) Program.* https://www.energy.gov/ne/light-water-reactor-sustainability-lwrs-program.

7 U.S. DOE (2021). *3 Advanced Reactor Systems to Watch by 2030.* https://www.energy.gov/ne/articles/3-advanced-reactor-systems-watch-2030.

8 GEN IV. Technology Systems. 2022.

9 U.S. DOE (2022). *DOE Selects Sodium-Cooled Fast Reactor Design for Versatile Test Reactor in Idaho.* https://www.energy.gov/ne/articles/doe-selects-sodium-cooled-fast-reactor-design-versatile-test-reactor-idaho.

10 U.S. DOE (2022). *Advanced Small Modular Reactors (SMRs).*

11 Idaho National Laboratory (2022). *Advanced Small Modular Reactors.* Idaho National Laboratory.

12 U.S. DOE (2020). *First U.S. Small Modular Boiling Water Reactor Under Development.* https://www.energy.gov/ne/articles/first-us-small-modular-boiling-water-reactor-under-development.

13 U.S. DOE (2021). *Fission and Fusion: What is the Difference?* https://www.energy.gov/ne/articles/fission-and-fusion-what-difference.

14 U.S. DOE (2022). *Fusion Nuclear Science and Technology.*

15 GAO (2021). *Carbon Capture and Storage: Actions Needed to Improve DOE Management of Demonstration Projects.*

16 Osman, A.I., Hefny, M., Abdel Maksoud, M.I.A. et al. (2021). Recent advances in carbon capture storage and utilisation technologies: a review. *Environmental Chemistry Letters* 19: 797–849.

17 National Energy Technology Laboratory (2021). *Carbon Dioxide Capture Approaches.* https://netl.doe.gov/research/coal/energy-systems/gasification/gasifipedia/capture-approaches.

18 Smith, K.H., Ashkanani, H.E., Morsi, B.I., and Siefert, N.S. (2022). Physical solvents and techno-economic analysis for pre-combustion CO_2 capture: a review. *International Journal of Greenhouse Gas Control* 118: 103694. https://doi.org/10.1016/j.ijggc.2022.103694.

19 NETL (2022). *Guidelines/Handbook For The Design of Modular Gasification Systems.* DOE/NETL-2022/3209.

20 Yadav, S. and Mondal, S.S. (2022). A review on the progress and prospects of oxy-fuel carbon capture and sequestration (CCS) technology. *Fuel* 308: 122057. https://doi.org/10.1016/j.fuel.2021.122057.

21 Bui, M., Adjiman, C.S., Bardow, A. et al. (2018). Carbon capture and storage (CCS): the way forward. *Energy & Environmental Science* 11: 1062–1176. https://doi.org/10.1039/C7EE02342A.

22 Ekström, C., Schwendig, F., Biede, O. et al. (2009). Techno-economic evaluations and benchmarking of pre-combustion CO_2 capture and oxy-fuel processes developed in the European ENCAP Project. *Energy Procedia* 1: 4233–4240. https://doi.org/10.1016/j.egypro.2009.02.234.

23 Global CCS Institute. Status Report 2023. https://www.globalccsinstitute.com/.

24 National Energy Technology Laboratory (2022). *Point Source Carbon Capture from Power Generation Sources.*

25 National Energy Technology Laboratory (2021). *Supercritical CO_2 power cycles for FE.* https://netl.doe.gov/node/7548.

26 National Energy Technology Laboratory (2021). *sCO2 Technology.* https://www.netl.doe.gov/carbon-management/sco2.

27 U.S. DOE (2020). *Supercritical CO_2 Technology.* https://www.energy.gov/supercritical-co2-tech-team.

28 National Energy Technology Laboratory (2021). https://netl.doe.gov/carbon-dioxide-removal. *Carbon Dioxide Removal.*

29 Valera-Medina, A., Xiao, H., Owen-Jones, M. et al. (2018). Ammonia for power. *Progress in Energy and Combustion Science* 69: 63–102. https://doi.org/10.1016/j.pecs.2018.07.001.

30 Palys, M.J. and Daoutidis, P. (2020). Using hydrogen and ammonia for renewable energy storage: a geographically comprehensive techno-economic study. *Computers & Chemical Engineering* 136: 106785.

31 Valera-Medina, A., Amer-Hatem, F., Azad, A.K. et al. (2021). Review on ammonia as a potential fuel: from synthesis to economics. *Energy Fuels* 35: 6964–7029. https://doi.org/10.1021/acs.energyfuels.0c03685.

32 Lipman T and Shah N (2007). Ammonia as an alternative energy storage medium for hydrogen fuel cells: Scientific and technical review for near-term stationary power demonstration projects, final report.

33 Afif, A., Radenahmad, N., Cheok, Q. et al. (2016). Ammonia-fed fuel cells: a comprehensive review. *Renewable and Sustainable Energy Reviews* 60: 822–835.

34 Zhao, Y., Setzler, B.P., Wang, J. et al. (2019). An efficient direct ammonia fuel cell for affordable carbon-neutral transportation. *Joule* 3: 2472–2484.

35 Rouwenhorst K.H.R. (2018). Power-to-ammonia-to-power (P2A2P) for local electricity storage in 2025: current developments, process proposal & future research required.

36 Clarke, L.E., Jiang, K., Akimoto, K. et al. (2014). Assessing transformation pathways. In: *Climate Change 2014: Mitigation of Climate Change. Contribution of Working Group III to the Fifth Assessment Report of the Intergovernmental*

Panel on Climate Change. Cambridge, United Kingdom and New York, NY, US: Cambridge University Press.

37 Minx, J.C., Lamb, W.F., Callaghan, M.W. et al. (2018). Negative emissions—Part 1: research landscape and synthesis. *Environment Research Letters* 13: 063001. https://doi.org/10.1088/1748-9326/aabf9b.

38 IPCC (2013). *Climate Change 2013: The Physical Science Basis.*

39 Caldecott, B., Lomax, G., and Workman, M. (2015). *Stranded Carbon Assets and Negative Emissions Technologies.* University of Oxford.

40 National Research Council (2015). *Climate Intervention: Carbon dioxide removal and reliable sequestration.* National Academies Press.

41 Lewis, E., McNaul, S., Jamieson, M. et al. (2022). Comparison of commercial, state-of-the-art, fossil-based hydrogen production technologies. In: *National Energy Technology Laboratory (NETL).* Pittsburgh, PA, Morgantown, WV, and Albany, OR (United States): https://doi.org/10.2172/1862910.

3

Power Storage

Power storage technologies can save energy and use it later where and when it is needed. Most renewable power generation technologies (e.g., solar, wind, and marine) vary based on weather conditions, time, location, and season. These challenges can be addressed with proper power storage technology. Existing commercialized power storage technologies can be classified into two main categories: batteries and mechanical energy storage. Battery technologies have unique features, e.g., size, shape, and portability. There are several different technologies for power storage, each with its own benefits and drawbacks (Figure 3.1). Also, chemical storage technologies (e.g., hydrogen and ammonia) have the potential to store energy and convert it to power in the future.

Several parameters play a crucial role in power storage, such as efficiency, lifespan (years), life cycle, energy density (Wh/L), capacity (MW or GW), safety, and cost. Efficiency measures how much power is lost or can be stored and recovered from the power storage system compared to the power used to charge and discharge the system. The life cycle measures how often the storage system can be adequately used before losing its ability to store power. Energy density estimates the amount of power that can be saved per volume (Wh/L). Higher efficiency means less power loss, and higher energy density means more energy storage in a smaller device. For storage systems, safety is critical, especially in high-capacity storage systems. For example, storage systems must be designed to prevent overheating and explosions, especially due to external temperature changes. Cost is another critical parameter in power storage, which determines commercialization viability, makes them more accessible and affordable, and increases widespread adoption. More detailed information has been provided by [1, 2].

This chapter presents an overview of mature low-emission power storage solutions. Particularly, it covers six technologies with high technology readiness level, along with a case study about pumped hydropower energy storage.

Net-Zero and Low Carbon Solutions for the Energy Sector: A Guide to Decarbonization Technologies,
First Edition. Amin Mirkouei.
© 2024 John Wiley & Sons, Inc. Published 2024 by John Wiley & Sons, Inc.

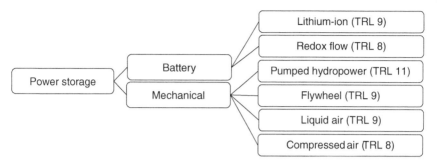

Figure 3.1   Classification of power storage technologies.

3.1   Battery

Batteries are electrochemical power storage systems with numerous benefits, such as being portable and easy to use, as well as high energy efficiency and density. The main drawbacks of batteries are short lifetime and high capital costs, along with high environmental impacts (e.g., hazardous waste) due to their raw materials, including metals or nonmetals. Earlier studies show that the grid-scale (MW) battery has significant environmental impacts. Currently, lithium-ion (Figure 3.2) and vanadium redox flow batteries have the highest TRL. Other types are sodium–sulfur (Na–S), nickel–cadmium (Ni–Cd), and lead–acid (Pb–A). Battery lifetime is 5–20 years, and their efficiency is around 65–90%.

Figure 3.2   Example of lithium-ion batteries. Source: Wiley Digital Library.

The basic science behind batteries involves the movement of charged particles, such as electrons or ions, between different materials or electrodes. The key parameters that affect the performance of batteries include: (i) electrode materials (e.g., lithium cobalt oxide, nickel–cadmium, and lead–acid) that greatly affect the performance of batteries, including its capacity, voltage, and lifespan; (ii) electrolyte (e.g., lithium salt solutions, sulfuric acid, and potassium hydroxide) that is the medium through which ions move between the electrodes, and it also affects the performance of batteries; (iii) separator, which is a permeable membrane that separates the two electrodes, allowing ions to pass through while preventing direct contact between the electrodes, and affects the capacity of batteries, internal resistance, and safety; (iv) temperature that affects the rate of chemical reactions inside the battery and can greatly impact its performance and increase the capacity of batteries but can also shorten its lifespan; (v) charging and discharging that affects their capacity, voltage, and lifespan; and (vi) battery management system that can monitor and control the battery charge and discharge rates, temperature, and other factors to improve its performance and lifespan. Overcharging, overdischarging, and rapid charging can all damage the battery and reduce its performance. Recent studies show that battery production requires 350–650 MJ/kWh energy, which releases 120–250 kg CO_2 eq. per kWh [1]. Other technologies, such as lithium iron phosphate batteries, have been explored due to their cost, safety, and durability. Pb–A batteries have a large market share (MWh) for rechargeable batteries for the automotive industry and telecommunications due to low production cost, wide size range, and efficiency in terms of charge retention and performance in different temperatures. The drawbacks are energy density, lifespan, and acid/lead leaks if breached.

3.1.1 Lithium-ion (TRL 9)

The basic science behind lithium-ion batteries involves electrochemical reactions, using lithium ions to generate power. Particularly, lithium ions move from the cathode to the anode during charging, where they are stored, and the ions return to the cathode during discharging, which creates a flow of electrons and generates power. The main components are two electrodes (cathode and anode), an electrolyte, and a separator. The primary materials used are lithium cobalt oxide (27%), steel (20%), graphite (16%), and polymer (14%) [1]. Most existing lithium-ion batteries are made of the positive cathode electrode from lithium cobalt oxide, the negative anode electrode from graphite, and the electrolyte from a lithium salt dissolved in an organic solvent, as well as the separator that is made of a thin, porous material to physically separate the cathode and anode while allowing lithium ions to move between the electrodes during charging and discharging.

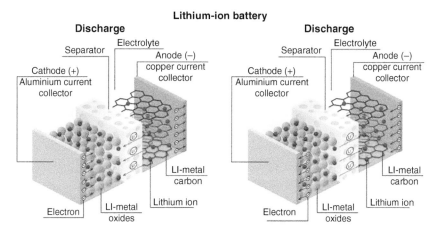

Figure 3.3 Schematic of lithium-ion power storage technology. Source: U.S. Department of Energy (public domain).

Currently, lithium-ion batteries are the most widely used energy storage system, covering over 90% of the market in 2020 (Figure 3.3) [3]. The key parameters to evaluate them are efficiency, energy density, and charge cycles (i.e., full charging and discharging). Lithium-ion batteries have high efficiency and energy density of up to 97% and 500 Wh/L, respectively. They can also last over 1,000 charge cycles. The main benefits of these batteries are: (i) high energy density to store more power in small size and weight, which makes them popular for use in portable devices and electric vehicles; (ii) long lifetime, depending on several factors, such as materials quality, applications, and charging and discharging conditions; and (iii) rapid charging and stability in various temperatures. The main drawbacks include: (i) capacity degradation over time due to natural chemical reactions and extreme temperature changes; (ii) environmental impacts due to toxic chemicals (e.g., lithium and cobalt) and high greenhouse gas (GHG) emissions from the production process; (iii) expensive and complex handling and recycling requirements; and (iv) safety issues, such as internal short circuits or fire. These batteries can also be used for microgrid scales. The major lithium-ion battery manufacturers in 2021 are Chinese companies, such as Contemporary Amperex Technology, BYD, and Ningde Amperex Technology. Other manufacturers are Samsung, LG, Panasonic, and Toshiba.

The significant environmental impacts associated with lithium-ion batteries include: (i) resource extraction, (ii) energy-intensive production process, (iii) limited recycling or disposal infrastructure, and (iv) safety concerns. For example, resources used (e.g., lithium, cobalt, and other metals) in these batteries are mainly mined in countries with limited environmental regulations.

Also, producing lithium-ion batteries requires more energy than other batteries (e.g., lead–acid) with a similar capacity. Other environmental impacts include properly recycling or disposing of lithium-ion batteries because they contain toxic chemicals and require expensive and complex recycling infrastructure. Currently, most of them end up in landfills, and the toxic chemicals leach into the soil and groundwater. There is limited infrastructure for recycling these batteries. Most batteries have been shipped to developing countries for processing, where they can be dismantled under unsafe conditions and create additional environmental and health hazards. While rare, incidents such as the Samsung Galaxy Note 7 recall have raised concerns about the safety of these batteries. Overall, lithium-ion batteries have become the dominant technology for portable electronic devices and electric vehicles due to their high energy density, long lifetime, and low self-discharge rate. However, significant environmental and social challenges associated with their production and disposal need to be addressed to ensure their long-term sustainability.

3.1.2 Redox Flow (TRL 8)

The basic science behind redox flow batteries involves electrochemical reactions that utilize liquid electrolytes that flow through the battery during charging and discharging. These batteries are made of two different electrolytes, separated by a membrane, to store and release energy. The electrolytes are derived from metal ions in various oxidation states (e.g., vanadium, iron, or zinc). The electrolytes are stored in separate tanks and pumped through the battery during operation. The oxidation state of the metal ions determines the battery charge state. The membrane allows the flow of ions, but it prevents the mixing of the two electrolytes.

The key parameters of redox flow batteries are electrolytes, membrane, efficiency, energy density, and durability. The electrolytes' characteristics and membrane types can improve efficiency and performance, as well as reduce degradation over time. Currently, redox flow batteries have lower efficiency than lithium-ion batteries. But they can scale up faster than other batteries to achieve better energy storage capacity. Furthermore, the durability of these batteries is highly dependent on the materials' quality and operating conditions. Figure 3.4 presents a single-cell redox flow battery, depicting electrolytes flowing from storage tanks through the serpentine flow field within the electrochemical cell [4].

The benefits of redox flow batteries include large-scale energy storage applications, stability, low degradation, and long lifetime compared to other batteries (Table 3.1). The drawbacks of redox flow batteries include: (i) low energy and power density (25–35 Wh/L), (ii) high capital cost compared to lithium-ion batteries; (iii) environmental impacts due to mining and processing high metal

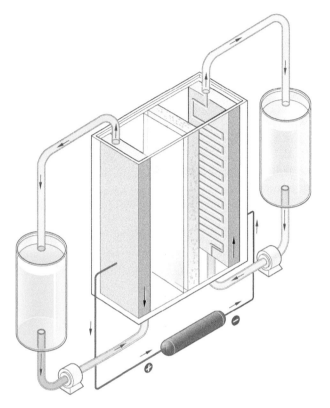

Figure 3.4 Schematic of redox flow batteries. Source: Small et al. [4].

requirements (e.g., vanadium, iron, and zinc); (iv) high energy requirements; and (v) large-scale infrastructure requirements, such as tanks for storing the electrolytes, pumps for circulating the electrolytes, and power electronics for controlling the flow of energy. These drawbacks can limit the adoption of redox

Table 3.1 Comparison of lithium-ion batteries with redox flow batteries.

Technology	Lithium-ion	Redox flow
Benefits	• High energy density (well-suited for applications with limited space) • High efficiency (well-suited for electric vehicles applications); less expensive	• High scalability (better fit for large-scale energy storage applications) • Longer life cycle (well-suited for frequent cycling, e.g., solar systems)

flow batteries in some applications and increase the capital cost to build the necessary infrastructure. The earlier techno-economic studies reported that the major cost drivers are materials, production processes, and the need for large-scale infrastructure. Currently, the redox flow battery has TRL 8 and a relatively small market. The main environmental impacts are from the following: (i) mining operations that contribute to soil erosion, water and air pollution, and habitat and land destruction; (ii) energy-intensive processes that contribute to GHG emissions; and (iii) toxic materials (e.g., vanadium and sulfuric acid) that can pose environmental risks. Currently, recycling redox flow batteries is not widely practiced due to its high cost and complexity.

In 2022, the main manufacturers of redox flow batteries were Sumitomo Electric Industries, UniEnergy Technologies, Gildemeister Energy Solutions, Redflow, Primus Power, Avalon Battery, Vionx Energy, and CellCube Energy Storage Systems. Vanadium redox is the most common one. Overall, redox flow batteries are a grid-scale power storage technology that can quickly provide a large amount of power. They offer different scales of energy storage applications due to their scalability, long lifetime, and potential for high efficiency.

3.2 Mechanical Energy Storage

Mechanical energy storage technologies use kinetic (energy of motion), weight, or gravitational forces for saving energy. The commercialized and most used mechanical energy storage systems are pumped hydropower, flywheel, liquid air, and compressed air energy storage. The main difference between batteries and mechanical energy storage technologies is their operational and scientific concepts (e.g., materials science, electrochemistry, and thermodynamics). Particularly, batteries work by utilizing electrochemical reactions, and mechanical energy storage technologies work by using kinetic energy for storing and releasing power.

The key parameters of energy storage systems are widespread adoption, efficiency, energy density, safety, and environmental impacts. The main barriers to mechanical energy storage systems are high capital costs due to large infrastructure and low efficiency. Compared to batteries, mechanical energy storage technologies do not require toxic chemicals (e.g., lithium and cobalt) that negatively impact the environment. They can be built from recycled steel and have a longer life with fewer replacements.

3.2.1 Pumped Hydropower (TRL 11)

Pumped hydropower energy storage technology can store energy and generate power from two water reservoirs at different elevations. The science behind pumped energy storage involves storing energy in the form of potential energy using water and gravity like a giant battery. Particularly, the energy is used to pump the water to a higher elevation (charge) and run the turbines when water moves down (discharge) to generate power [5]. This technology has been used since 1890 in Europe and 1930 in the United States, and its TRL is 11, meaning it has been commercialized worldwide and reached the highest proof of stability and growth.

According to the U.S. Department of Energy (DOE), the pumped hydropower technology was used in 43 plants and stored over 93% of utility-scale energy in the United States in 2021. Currently, there are two main types of pumped storage: open-loop and closed-loop (Figure 3.5). The main difference is the connection to the natural body of water (e.g., rivers or lakes). The closed-loop reservoirs are not connected to the outside or natural water, and most of the new plants in the United States are closed-loop. The main challenge is to find locations (topographies) with significations (low and high) in close proximity. There are other types, such as pumping water into tanks full of gas or air to compress and pressurize the gas and open it to push the water and run the turbine when the grid needs power.

The key parameters that affect pumped energy storage technology are efficiency, capacity, elevation, location, infrastructure costs, and environmental impacts. Highly efficient energy storage systems can reduce energy loss during storage and retrieval. Higher elevation and capacity can store more potential energy. In addition, infrastructure and maintenance costs and suitable location are critical

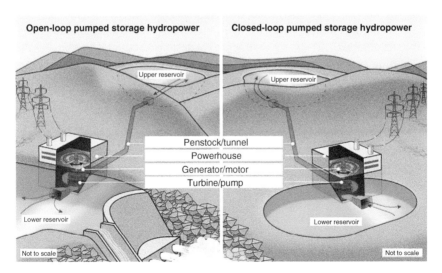

Figure 3.5　Schematic of pumped storage. Source: U.S. Department of Energy (public domain).

regarding feasibility with specific geographical features and accessibility to main power grids. The earlier techno-economic studies reported that the major cost drivers include building reservoirs, tunnels, and powerhouses. The significant environmental impacts of this technology include the adverse effects on wildlife, domestic ecosystem, and water resources. For example, there can be loss of habitat, aquatic species, and ecosystems due to flooding large areas and changing the water flow and temperature. Other impacts include GHG emissions from facility construction and land use changes, affecting local communities, such as agricultural farms.

Currently, the top five countries that use pumped hydropower energy storage technology are China, the United States, Japan, Switzerland, and Germany, with over 30, 22, 7, 1.8, and 1.5 GW of installed capacity, respectively. In 2022, the largest plant was the Three Gorges Dam in China, with over 22 GW of capacity. Currently, the largest plant in the United States is Bath County Pumped Storage Station in Virginia, with around 3 GW of capacity. Overall, the main advantages of pumped hydropower storage technology are its high capacity, low cost per kWh, long life-time, and reliability. The disadvantages are geographical limitations, high capital costs, and environmental concerns [5].

3.2.2 Flywheel (TRL 9)

Flywheel is one of the earliest energy storage technologies, using a rotating mechanical device with widespread applications (e.g., aerospace and telecommunications) to store energy (MW) and discharge in a few minutes (Figure 3.6). The science behind flywheel energy storage systems involves storing kinetic energy in rotational motion. The key scientific parameters are rotational motion, inertia, and friction or losses from air resistance or bearing friction. The main factors that can impact the performance of flywheel energy storage systems are rotor

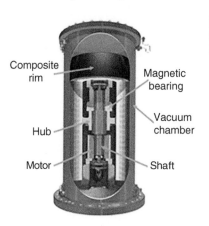

Figure 3.6 Schematic of flywheel power storage technologies. Source: Beacon Power LLC.

Composite rim

Magnetic bearing

Vacuum chamber

Hub

Motor

Shaft

mass, rotor speed, bearing friction, and materials. Higher rotor mass or speed can increase the amount of stored energy. The bearing friction from supporting the heavy rotor (wheel) can reduce the efficiency and durability of the system by losing power.

The main components of flywheel energy storage technology are: (i) a vacuum chamber for reducing the air resistance and maximizing the efficiency of the flywheel, (ii) a cylinder with a large mass that spins at a substantial speed of several thousand RPM, (iii) a motor for storing energy by spinning the flywheel and recovering the stored energy by spinning back and acting as a generator, (iv) magnetic bearings for supporting the flywheel and reducing the mechanical wear and frictional losses. The rotating cylinder can be made out of steel or carbon composite. The composite is lighter and stronger and can achieve higher rotational speeds.

The benefits of this technology are low maintenance, long lifetime, high power density, fast response times, and low environmental impacts during operation. The main drawbacks are high capital cost, limited energy storage duration (seconds to minutes), and the need for sophisticated control systems because mechanical failure can be catastrophic due to massive forces or overheating. The earlier techno-economic studies reported that the major cost drivers are manufacturing, installation, and maintenance. Currently, the flywheel energy storage has TRL 9, but the market for this energy storage technology is relatively small compared to other technologies. The main environmental impacts are material extraction and processing, such as steel, copper, and rare earth metals. Other impacts include high noise and vibration during the operation. Currently, the major flywheel energy storage systems manufacturers in the United States are Beacon Power, Vycon, Pentadyne Power, and Active Power, with up to 1, 2.5, 3, and 4 MW storage capacity, respectively. Overall, flywheel energy storage technology is more appropriate for short-duration, high-power applications.

3.2.3 Liquid Air (TRL 9)

The basic science behind liquid air energy storage systems involves compression, cooling, and energy release. Liquid energy storage liquefies the air at very low temperatures at around $-200\,°C$ (cryogenic), stores it in a tank, then brings it back to the gas state with ambient air or waste heat from other industrial processes, and uses the gas to turn a turbine and generate power. The key parameters that affect the performance of these storage systems include process efficiency, thermal insulation, turbine design, and size. Improving these parameters reduces energy loss and increases performance. The benefits are a long lifetime, large-scale storage capacity (GWh), power outputs (100 MW), low environmental impacts, and fewer safety challenges. The main drawbacks are the high cost of manufacturing and installation, and limited energy storage duration (hours to days).

The techno-economic studies show that large-scale systems are more cost-effective than smaller ones. Currently, the liquid air energy storage has TRL 9, but the market is new and growing. The leading liquid air energy storage manufacturers are Highview Power, Cryostar, Linde, and Storelectric, with up to 200 MW, 1.2 GW, 100 MW, and 50 MWh power storage capacity, respectively. More details about this technology have been provided by [6].

3.2.4 Compressed Air (TRL 8)

The basic science of compressed air energy storage is similar to liquid air energy storage, involving compression, storage, and energy release. It uses ambient air or another gas to compress and store energy under high pressure (around 1,000 psi) in underground containers (Figure 3.7). It is very similar to pumped storage in terms of plant size and applications, and has the potential for both small- and large-scale storage needs. The main methods for compressed air energy storage are diabatic or adiabatic. In adiabatic, the heat from air compression is stored, recovered, and reused to reheat the compressed air and increase efficiency by up to 70%.

The key parameters that affect the efficiency and performance of compressed air energy storage are capacity, turbine design, and air quality. The main benefits of compressed air energy storage systems are high capacity, low cost per kWh, low storage losses, and long storage periods (days). The issues are high capital cost,

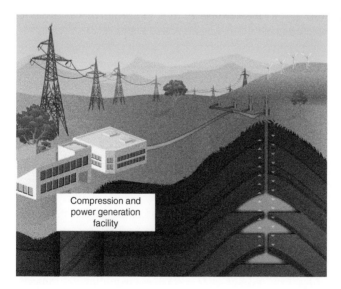

Compression and power generation facility

Figure 3.7 Schematic of compressed air power storage technologies. Source: Pacific Northwest National Laboratory [7].

Table 3.2 Comparison of different energy storage technologies.

Technology	Medium	Capacity Range	Type	Duration range	Potential Environmental Impacts
Commercialized batteries	Lithium, nickel, and cobalt	50 kW–1 MW	Electrochemical	Hours to days	Carbon emissions and toxic waste
Advanced batteries	Lithium and others	1 MW–10 MW	Electrochemical	Hours to days	Rare material use and toxic waste
Chemical storage	Ammonia, hydrogen, methane, and methanol	50 kW–1 GW	Chemical	Days to months	Air pollution and chemical waste
Supercapacitors	Electrolyte	60 kW–5 MW	Electrical	Seconds	Material use and hazardous waste
Flywheels	Rotating mass	50 kW–1 MW	Mechanical	Seconds to minutes	Noise pollution, energy, and material use
Pumped hydropower	Water	100 MW–1 GW	Mechanical	Days to weeks	Water resources, land use, and ecosystems
Thermochemical (compressed air, electrothermal)	Air	10 MW–1 GW	Thermo-mechanical	Hours to days	Air and noise pollution, land and energy use

Source: Adapted from Valera-Medina et al. [8].

low energy density, and site selection, which requires large and suitable geology to store compressed air underground. Currently, this technology has TRL 8 and a relatively small market. The leading compressed air energy storage system manufacturers are General Electric, Dresser-Rand (a subsidiary of Siemens Energy), Apex CAES, and LightSail Energy. The main environmental impacts are high land use, air pollution, noise, and vibration during the operation. Overall, this technology can store large amounts of energy for long periods of time during off-peak (low demand) and provide power during peak load (high demand).

Overall, energy storage technologies have several benefits and drawbacks, depending on the application and operational requirements, such as storage duration, efficiency, cost, and environmental impacts. Table 3.2 compares the leading power storage technologies.

3.3 Case Study: Pumped Hydropower Energy Storage

A pumped hydropower system is one of the leading energy storage technologies in the world. Currently, China and the United States have the largest pumped hydropower systems. This technology uses excess power from the local power grid to pump water from the lower to upper reservoir during low-demand periods. Then, the water can be released from the upper reservoir to run the turbine and generate power during high demand. A pumped hydropower system is a reliable and flexible technology for providing power during high demand. Although pumped hydropower systems faced several social and environmental concerns, they became one of the vital energy storage technologies in the world because they can quickly stabilize regional power needs, respond to demand changes, and prevent blackouts. However, major droughts can reduce the capacity and limit the ability to recharge the reservoirs. Fengning Pumped Storage Power Station in Hebei Province is the world's largest pumped hydropower energy storage facility, operating since 2020 in China. The height difference between the two reservoirs is 535 m, with a 3.6 GW total installed capacity and 18 GWh storage capacity.

The United States has 42 pumped hydropower energy storage plants, with around 22.9 GW installed storage capacity [5]. Bath County Pumped Storage Station in Virginia is the world's second largest pumped hydropower energy storage facility [9], operating since 1985. It has two reservoirs with 384 m height difference between them, six generators, a 3 GW total installed capacity, and a 24 GWh storage capacity. For power generation, the water level in the upper reservoir can drop over 32 m, and the lower reservoir can rise around 18 m. The water flow for charging (pumping uphill during nights) is 12.7 million gallons per minute, and the water flow for discharging (pumping downhill during high demand) and power generation is approximately 13.5 million gallons per minute [10]. Earlier studies reported that Bath County Pumped Storage Station consumes 5 kW power to generate 4 kW power, meaning its efficiency is 80% with no GHG emissions since it does not use fossil fuels [11].

Overall, pumped hydropower energy storage plants serve as the largest power storage source or a massive battery for places with unsteady power demand. Otherwise, there is no reason to build these energy storage systems if the power demand is steady throughout the day. Also, pumped storage can start generating a significant amount of power in a few minutes without downtime or refueling.

Self-Check Questions

1. What are the main types of power storage technologies?
2. What are the critical parameters in power storage technologies?
3. What are the benefits and drawbacks of batteries for storing power?
4. What are the primary commercialized batteries?
5. What are the key parameters that affect the performance of batteries?
6. How does a lithium-ion battery work?
7. What are the main components of lithium-ion batteries?
8. What are the benefits of lithium-ion batteries?
9. What are the drawbacks of lithium-ion batteries?
10. What are the main environmental impacts associated with lithium-ion batteries?
11. How does a redox flow battery work? And what are the main components of these batteries?
12. What are the key parameters of redox flow batteries?
13. What are the benefits of redox flow batteries?
14. What are the drawbacks of redox flow batteries?
15. What are the main environmental impacts associated with redox flow batteries?
16. How does mechanical energy storage work?
17. What are the most used mechanical energy storage systems?
18. What are the differences between batteries and mechanical energy storage systems?
19. How does pumped hydropower store energy?
20. What are the main types of pumped hydropower energy storage systems? And what are the main differences?
21. What are the key parameters that affect pumped hydropower energy storage systems?
22. What are the advantages and disadvantages of pumped hydropower energy storage systems?

23. How does flywheel energy storage work? And what are the main parameters of this technology?
24. What are the main factors affecting the performance of flywheel energy storage technology?
25. What are the main components of flywheel energy storage technology?
26. What are the benefits and drawbacks of flywheel energy storage technology?
27. What are the main environmental impacts of flywheel energy storage technology?
28. How does liquid air energy storage work?
29. What are the main parameters of liquid air energy storage technology?
30. What are the benefits and drawbacks of liquid air energy storage technology?
31. How does compressed air energy storage work?
32. What are the key parameters that affect the efficiency and performance of compressed air energy storage?
33. What are the benefits and issues of compressed air energy storage systems?
34. What are the main environmental impacts of compressed air energy storage systems?

References

1 Dehghani-Sanij, A.R., Tharumalingam, E., Dusseault, M.B., and Fraser, R. (2019). Study of energy storage systems and environmental challenges of batteries. *Renewable and Sustainable Energy Reviews* 104: 192–208. https://doi.org/10.1016/j.rser.2019.01.023.

2 Lazard (2020). *Levelized Cost of Energy and of Storage* 2022.

3 IEA (2022). Energy Storage.

4 Small, L.J., Fujimoto, C.H., and Anderson, T.M. (2020). *Redox Flow Batteries*, US DOE Energy Storage Handbook.

5 US DOE (2022). *Pumped Storage Hydropower: A Key Part of Our Clean Energy Future.*

6 Vecchi, A., Li, Y., Ding, Y. et al. (2021). Liquid air energy storage (LAES): a review on technology state-of-the-art, integration pathways and future perspectives. *Advances in Applied Energy* 3: 100047. https://doi.org/10.1016/j.adapen.2021.100047.

7 PNNL (2019). Compressed Air Energy Storage 2019. https://caes.pnnl.gov/.

8 Valera-Medina, A., Amer-Hatem, F., Azad, A.K. et al. (2021). Review on ammonia as a potential fuel: from synthesis to economics. *Energy Fuels* 35: 6964–7029. https://doi.org/10.1021/acs.energyfuels.0c03685.

9 EIA (2022). Independent Statistics and Analysis 2022. https://www.eia.gov/state/analysis.php?sid=VA#_ftn1 (accessed March 28, 2023).

10 Dominion Energy (2022). Bath County Pumped Storage Station 2022. https://www.dominionenergy.com/projects-and-facilities/hydroelectric-power-facilities-and-projects/bath-county-pumped-storage-station (accessed March 28, 2023).

11 Virginia Places (n.d). Pumped Storage in Bath County, http://www.virginiaplaces.org/energy/bathpumped.html#three (accessed March 28, 2023).

4

Heat Generation

The basic science behind heat generation involves heat transfer mechanisms, including conduction, convection, or radiation. The key parameters are the heat sources (e.g., fossil fuels, geothermal, or solar radiation) and the input–output energy ratio. The heat demand in commercial and residential buildings accounts for over 60% of the total energy needed in cold regions and over 30% in warm regions [1]. The major industrial applications of heat generation are metal processing, distract heating, crude oil heating, food processing, pharmaceutical, and chemical industries. Heat generation technologies highly depend on the heat source (i.e., renewable or non-renewable).

Non-renewable (fossil fuel)-based heat generation technologies (e.g., natural gas-fired or coal-fired boilers and furnaces) have been widely used due to their high energy density; however, they release high greenhouse gas (GHG) emissions. Globally, heating generation is one of the major sources of GHG emissions, using mainly fossil fuel-based (e.g., coal, oil, and natural gas) boilers. The main environmental impacts include air pollution from fossil fuel combustion, water and soil pollution from waste discharge, and ecosystem disruption due to resource extraction. Recent studies reported that heat generation technologies could reduce environmental impacts by improving efficiency and using renewable sources (e.g., solar or geothermal). The primary renewable sources for heat generation are solar thermal, geothermal, and biomass.

This chapter focuses on low-emission, mature heat generation technologies: solar thermal district heating systems and large-scale heat pumps. The large-scale heat pump is an efficient technology that can reduce energy consumption and GHG emissions through heat waste recovery, process integration, and renewable energy consumption. The leading producers of heat generation technologies are fossil fuel companies (e.g., Shell and ExxonMobil) and renewable energy companies, such as GE and Siemens.

Net-Zero and Low Carbon Solutions for the Energy Sector: A Guide to Decarbonization Technologies,
First Edition. Amin Mirkouei.
© 2024 John Wiley & Sons, Inc. Published 2024 by John Wiley & Sons, Inc.

4.1   Low-Emission Technologies

Low-emission heat generation technologies can minimize the environmental impacts by utilizing renewable sources, but they may require higher capital costs due to specific design and infrastructure.

4.1.1   Solar Thermal District Heating (TRL 10)

Solar thermal district heating technology uses solar-based energy to generate and distribute heat to a network of buildings. Solar-based energy can be generated everywhere, especially in deserts, remote, or low-income regions. The basic science behind solar thermal district heating technology involves capturing solar radiation with solar collectors to heat up the transfer fluid and then circulating the heat through the system. The main components of this technology include solar collectors, heat storage tanks, heat exchangers, and distribution pipes.

This technology can play a key role for those regions that need high amounts of gas or oil for addressing heat demands. For example, Canada and Nordic countries have high heating demands per capita worldwide. They also set ambitious targets to become 100% renewable energy country and society by 2050 to reduce GHG emissions for the coming decades. The increasing interest in solar thermal energy for satisfying heat generation has been evident in Nordic countries during the last decade. Some plants are mixed and operated with natural gas boilers to meet the power and heat needs during the cold seasons, such as winter with low sunshine. The district heating method can efficiently generate and distribute heat commercially and recycle heat waste. It has been in use since the fourteenth century in Europe, utilizing geothermally heated water [2].

Solar thermal district heating has been fully commercialized to supply heat, mainly in Europe (e.g., Denmark and Finland). This technology can be classified into two types: (i) centralized systems with a central solar collector and storage and (ii) decentralized systems with distributed solar collectors (e.g., roof space) and without storage (Figure 4.1). Centralized systems with ground-mounted solar collectors and seasonal storage have been fully commercialized, but decentralized systems have not been commercialized yet. Solar hot water is one of the most effective, clean methods to use the sun's radiation and warm up the water for showers, washing dishes and clothes, and laundry. This solution can reduce over 50% of energy consumption for water heating and cut over 4 gigatons of GHG by 2050. The payback periods are between 2 and 4 years, depending on the location and system capacity. Concentrated solar thermal systems can generate heat and steam to run the turbines in dry and hot regions. This solution can generate up to 7% of world energy by 2050 and cut roughly 24 gigatons of GHGs. The leading solar thermal district heating technology manufacturers are Arcon-Sunmark,

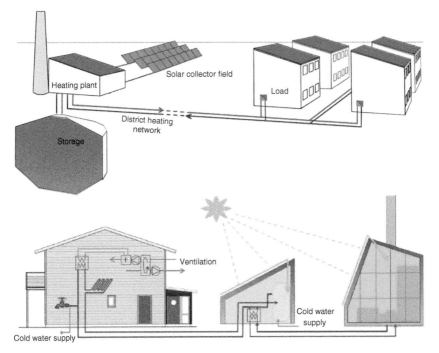

Figure 4.1 Schematic of centralized (top) and decentralized (bottom) solar thermal district heating system. Source: Perez-Mora et al. [2]/John Wiley & Sons.

ABB, and Thermax. Further details about solar thermal district heating have been provided by [2].

4.1.2 Large-Scale Heat Pumps (TRL 9)

The basic science behind heat pumps involves generating or transferring heat from a low-temperature source to a high-temperature sink. They use the thermodynamic cycle principle to warm or cool the house and provide a comfortable temperature. The main types are air-source or ground (geothermal)-source heat pumps that can provide up to 160 °C hot air and steam for various process needs (Figure 4.2). Heat pump technologies can be used for various applications, such as heating and cooling for buildings or industrial purposes. Large-scale heat pumps have been recognized as heat decarbonization technology due to high energy utilization efficiency in recovering the heat waste from industrial processes and eventually replacing other energy-intensive methods (e.g., fossil fuel-based boilers) by using renewable energy sources.

Large-scale heat pumps can be classified based on their temperature level and refrigerant, such as standard (up to 80 °C), high (up to 100 °C), or very high

(a)

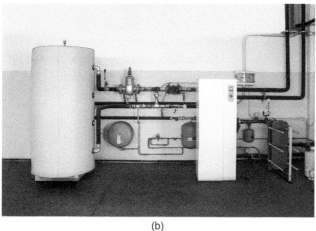

(b)

Figure 4.2 Ducted air-source heat pumps (a) and geothermal heat pumps (b).
Source: U.S. Department of Energy (public domain) [3].

(up to 160 °C) temperature heat pumps with synthetic–organic refrigerants (e.g., hydrochlorofluorocarbon, hydrofluorocarbon, or hydrofluoroolefin) or with R717 refrigerant. Refrigerants transfer heat through compressing, condensing, and then expanding. The main components of large-scale heat pumps are the compressor, evaporator, condenser, and expansion units. The key parameters are the heat pump's efficiency (the ratio of heat output to energy input), capacity, and temperature range. Their main benefits are a long lifetime, low operational costs, low maintenance, low emissions, and high safety compared to other conventional heat generation technologies, such as coal-fired boilers. However, their drawbacks are high capital costs for the main components and potential refrigerant leaks. Large-scale heat pumps reached TRL 9, and the leading producers are Carrier, Trane, Johnson Control, Danfoss, and Mitsubishi Electric. More detailed information about this technology has been provided by Schlosser et al. [4] and Jiang et al. [5].

4.2 Case Study: Geothermal Heat Pumps

Geothermal can be an energy source for heating, cooling, and power generation. The basic science behind geothermal heat pumps involves using the constant Earth's temperature throughout the year because the temperature at about 10 m below the surface remains relatively stable between 10 and 15 °C. Figure 4.3 presents community-scale heating and cooling system using geothermal bore-holes up to 150 m deep. Geothermal heat pumps can take advantage of the Earth as a heat source and sink (thermal storage) to efficiently exchange temperatures from underground for heating and cooling in winter and summer. The main components are the heat pump, underground loop (series of pipes), and fluid that can circulate in the loop by absorbing heat from the ground for heating or absorbing heat from the building for cooling. The loop can be developed either horizontally or vertically.

Similar to large-scale heat pumps, the main benefits are high efficiency, low operational costs, low energy use (up to 60%), and low emissions. The drawbacks are high capital costs and high land use. Some of the leading geothermal heat pump producers are Trane, Carrier, Bosch, and WaterFurnace. Table 4.1 presents some of the main geothermal heat pump project in the world.

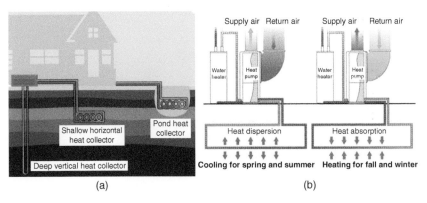

Figure 4.3 Schematic of geothermal heat pumps: (a) different collector types and connection used and (b) heating and cooling systems. Source: U.S. Department of Energy (public domain).

Table 4.1 Examples of geothermal heat pump projects in the world.

Project	Location	Thermal Energy Capacity (MW)	Features	Reference
Ball State University	USA	17	3,400 boreholes drilled between 120 and 150 m for providing heating and cooling to 47 buildings	[8]
Boden Ice Arena	Sweden	2.5	2,700 m of boreholes drilled 240 m for providing heating and cooling spaces	[9]
Drake Landing Solar Community	Canada	1.5	144 boreholes drilled 37 m deep for providing space heating and hot water for 52 homes	[10]

Self-Check Questions

1. What is the basic science behind heat generation?
2. What are the key parameters and the primary demand for heat generation?
3. What are the major applications and needs for heat generation?
4. What are the main non-renewable (fossil fuel)-based heat generation technologies? And what are their advantages and disadvantages?
5. What are the main renewable resources for heat generation technologies? And what are their pros and cons?
6. What is solar thermal district heating technology? And how does it work?
7. What are the main components of solar thermal district heating technology?
8. What are the main types of solar thermal district heating technology?
9. What is the science behind heat generation through heat pumps?
10. What are the main types and applications of heat pumps?
11. What are the main types of large-scale heat pumps?
12. What are the main components of large-scale heat pumps?
13. What are the key parameters of large-scale heat pumps?
14. What are the benefits and drawbacks of large-scale heat pumps?
15. How does a geothermal heat pump work?
16. What are the main components of the geothermal heat pump for heat generation?
17. What are the benefits and drawbacks of geothermal heat pumps?

References

1 IRENA (2015). Solar Heating and Cooling for Residential Applications.

2 Perez-Mora, N., Bava, F., Andersen, M. et al. (2018). Solar district heating and cooling: a review. *International Journal of Energy Research* 42: 1419–1441.

3 U.S. DOE (2022). Heat Pump Systems. https://www.energy.gov/energysaver/heat-pump-systems.

4 Schlosser, F., Jesper, M., Vogelsang, J. et al. (2020). Large-scale heat pumps: applications, performance, economic feasibility and industrial integration. *Renewable and Sustainable Energy Reviews* 133: 110219. https://doi.org/10.1016/j.rser.2020.110219.

5 Jiang, J., Hu, B., Wang, R.Z. et al. (2022). A review and perspective on industry high-temperature heat pumps. *Renewable and Sustainable Energy Reviews* 161: 112106. https://doi.org/10.1016/j.rser.2022.112106.

6 U.S. DOE (2020). Geothermal Heat Pumps: Overview and Examples.

7 U.S. DOE (2022). Geothermal Heat Pumps. EnergyGov 2022. https://www.energy.gov/eere/geothermal/geothermal-heat-pumps (accessed July 23, 2023).

8 Ball State University (2012). Geothermal Case Studies 2012. https://eri.iu.edu/erit/case-studies/ball-state-university-geothermal.html (accessed March 30, 2023).

9 Energimyndigheten (2022). Sweden's energy situation 2022. https://www.energimyndigheten.se/ (accessed March 30, 2023).

10 Drake Landing Solar Community (2007). The District Heating System 2007. https://www.dlsc.ca/district.htm (accessed March 30, 2023).

5

Heat Storage

The energy transition aims to decarbonize the global energy systems by switching to renewable or low-carbon energy sources, which requires numerous innovations, adaptations, and new policies. The main drawback of some renewable energy sources (e.g., solar and wind) is the time limitation to generate energy throughout the day and night. Therefore, energy storage systems are necessary for storing and steadily supplying energy when needed. There are two main types of energy storage systems: (i) power storage, such as batteries for small scale or mechanical storage methods for large scale, and (ii) heat or thermal energy storage that can store energy from the sun, geothermal, or heat waste. Some heat storage technologies have been used on a large scale, such as latent, sensible, and thermochemical heat storage (Figure 5.1). The main parameters to compare different heat storage technologies are storage energy density, efficiency, heat capacity, storage period, cost, safety, and temperature range. These technologies can manage high power and store energy over a long period of time. More detailed information about heat storage systems has been provided by [1, 2].

An overview of key low-emission solutions for heat storage is presented in this chapter. Particularly, this chapter provides four technologies with high technology readiness level, along with a case study about commercial concentrating solar power (CSP) plants.

5.1   Physical

5.1.1   Sensible Heat Storage (TRL 9)

Sensible heat storage is based on increasing the temperature of an element or material (charging) and recovering the energy by dropping its temperature (discharging). Water is the widely used material for low temperatures (below 100 °C) due to its availability and nontoxic characteristics. For high temperatures (above

Net-Zero and Low Carbon Solutions for the Energy Sector: A Guide to Decarbonization Technologies, First Edition. Amin Mirkouei.

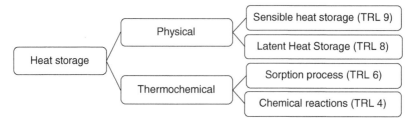

Figure 5.1 Classification of heat storage technologies.

Table 5.1 Comparison of sensible storage media properties.

Medium	Type	Storage Density (kJ/kg)	Temperature Range (°C)	Specific Heat (kJ/kg.k)
Aluminum oxide	Solid	455	200–700	1.3
Carbonate salts	Liquid	630	450–850	1.8
Hydroxide salts (e.g., NaOH)	Liquid	735	350–1,100	2.1
Graphite	Solid	665	500–850	1.9
Sodium chloride (NaCl)	Solid	315	200–500	0.9
Nitrate salts	Liquid	560	300–600	1.6
Sodium liquid metal	Liquid	455	316–700	1.3

100 °C), solid materials can be used for heat storage, such as ceramics and concrete. Molten salt has been widely used for high-temperature heat storage systems due to its low costs, availability, and chemical characteristics and attributes, such as high heat capacity, temperature stability, boiling point, and nonflammability. Molten salt can act as heat transfer and storage material, but the drawbacks are its high melting point, viscosity, and low thermal conductivity. Table 5.1 compares the thermophysical properties of sensible heat storage media [3].

Sensible heat storage technology has small energy density storage (15–50 kWh/m³), medium/high efficiency (50–90%), and small capacity (10–50 kWh/t). The technology is fully commercialized on the industrial scale mainly due to low capital cost compared to other heat storage methods. However, it requires isolation systems to prevent thermal losses over time [4]. The main parameters of sensible heat storage are material mass, specific heat capacity, and temperature range. Specific heat capacity is the amount of heat required to increase the temperature of a material (e.g., solid or liquid) to a certain amount. The main components are the storage tank, medium, and heat exchanger. The main benefits of sensible heat storage are high efficiency, low cost, no emissions during the operation, and a wide range of applications. However, sensible heat storage systems require large storage tanks due to lower energy density compared to other technologies. More details about sensible heat storage have been provided by [3].

5.1.2 Latent Heat Storage (TRL 8)

Latent heat storage is based on the amount of energy (heat) needed to change the material phase, for example, from solid to liquid or liquid to gas. The temperature range is between 20 and 80 °C, and the widely used phase-change materials (PCMs) are organic compounds and inorganic salts, such as water, molten salts, or sodium hydroxide. During the phase change, the heat forms or breaks the bond between the molecules that can absorb or release a large amount of heat. The main components are the storage medium or PCM and heat exchanger for transferring heat to the medium (Figure 5.2).

Some of the benefits of latent heat storage technology include: (i) storing and releasing heat at constant temperatures; (ii) high energy density storage (50–100 kWh/m^3), which requires smaller storage volumes; (iii) high efficiency (up to 90%); and (iv) medium capacity (50–100 kWh/t). This technology has various applications, especially for storing heat for several hours or days. The main drawbacks of this technology include: (i) low heat transfer rate (or thermal conductivity), (ii) high cost, (iii) limited storage periods, (iv) thermal losses, and (v) large temperature difference requirements between the storage material and heat source. Storage medium is one of the major cost drivers of this technology. Also, using fossil fuel-based mediums (e.g., paraffin wax) has negative environmental impacts from manufacturing and disposing of them. Renewable mediums (e.g., salt hydrates) have fewer environmental impacts. The leading producers of

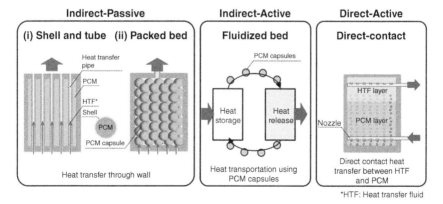

Figure 5.2 Schematics of different latent heat storage systems. Source: Nomura and Akiyama [5]/John Wiley & Sons.

latent heat storage systems are Calmac, Ice Energy, Viking Cold Solutions, and Sunamp. More details about latent heat storage and different storage media have been provided by [3, 6].

Aqueous salt solution is a type of latent heat storage that works by heating the storage medium (a mixture of water and salt) to a high temperature. The solution can reduce the freezing point and remain liquid at high temperatures. Then it can be used to store and release heat when needed to generate steam and power for various applications. The main components of this technology are the storage medium, tank, and heat exchanger. The salt concentration in the solution is one of the leading performance factors. The benefits of this technology are high efficiency, long lifetime, low cost, and high applications. The main drawback is the high-temperature requirements for the highly efficient operation. This technology has TRL 8-9 and is fully commercialized due to low environmental impacts and costs. Some of the leading manufacturers of aqueous salt solution systems for heat storage are SolarReserve, BrightSource Energy, and Abengoa.

5.2 Thermochemical

Thermochemical heat storage is based on reversible reactions (endothermic and exothermic) that can be either physical or chemical phenomena (Figure 5.3). During the endothermic reaction (charging), an element decomposes into two elements, using heat and storage separately. During the exothermic reaction (discharging), the heat is released from bonding the elements. Thermochemical heat storage can be an open or closed process with either integrated or separated storage [8]. This technology has high energy density storage (100–700 kWh/m^3), high efficiency (up to 99%), and high capacity (120–250 kWh/t). Due to its complexity and high capital cost, this technology is still in laboratory prototypes

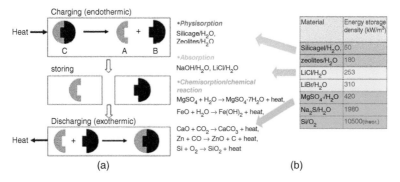

(a)   (b)

Figure 5.3 Schematic of thermochemical energy storage principles. Source: Dimitriev et al. [7]/John Wiley & Sons.

or pilot scale. There are two main thermochemical heat storage: chemical reactions and sorption-based processes. The main benefits of thermochemical heat storage technologies are high energy storage density and low energy losses over a long storage period, which make them well suited for long-term energy storage compared to latent and sensible heat storage systems.

5.2.1 Sorption Process (TRL 6)

Sorption-based heat storage uses thermochemical principles for storing energy, which are endothermic and exothermic or reversible chemical reactions to store and release thermal energy (Figure 5.4). The key components are sorbent materials, the reactor and heat exchanger, a vapor transport system, and an energy storage and retrieval system. The two main types of thermochemical sorption systems are adsorption and absorption, using sorbent materials with specific chemical compositions (e.g., water). Adsorption is a surface interaction between a gas and a solid. Particularly, gas is adsorbed on the sorbent materials that cause chemical reactions and release heat or thermal energy. The process can be repeated several times and release and store heat during adsorption and desorption, respectively. The desorption process includes heating or lowering the pressure to release gas or liquid that causes sorbent material to cool down and store heat. Absorption is the volume interaction between the gas or liquid (e.g., water) and another liquid (sorbent materials), such as salts and clays.

The main parameters that can affect the performance are process conditions (e.g., temperature and pressure), cycle time, and sorbent materials. This technology gained attention due to its low heat loss and high energy storage

Figure 5.4 Schematic of sorption-based heat storage system. Source: Henninger et al. [9]/John Wiley & Sons.

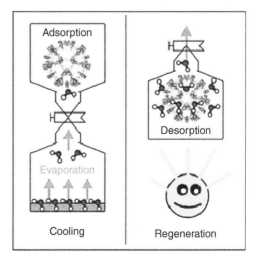

density, as well as its broad applications for space heating, hot water supply, short- and long-term energy storage, and industrial heat and cooling systems. Additionally, sorption-based heat storage systems do not release any emissions during the operation and have low environmental impacts from the production and disposal of the required resources. Some drawbacks are high cost, complex design, and sorbent material degradation over time. Also, sorbent materials can be toxic and hazardous and need special processes for production and disposal. Prior studies investigated the use of solar systems, reactor designs for a variety of reactions (e.g., solid–gas or gas–gas), and different mechanisms and materials (e.g., salt hydrates [10]) to check their thermal properties and sorption characteristics in sorption-based heating and cooling systems.

Sorption storage systems have TRL 6 and a relatively small market compared to other energy storage technologies. There are several companies that can offer a range of sorption storage systems (e.g., Chromasun and SorTech AG) for various applications. More detailed information about sorption-based thermochemical heat storage has been provided by [1].

5.2.2 Chemical Reactions (TRL 4)

Chemical-based heat storage systems can store thermal energy using chemical substances and reversible chemical reactions (store and release). The science behind chemical reactions involves forming and breaking chemical bonds to release and store heat in the form of chemical potential energy. The key components of chemical-based heat storage are the reactant, reactor, heat exchange, and energy storage system. The main factors on the performance are energy storage capacity, depending on the specific chemical reaction, and the energy balance depending on the heat input and output. Other key factors are reactant type, reactor temperature, and time.

Chemical-based heat storage can be classified into three main types based on the reaction mechanism, which are gas–gas, liquid–gas, or solid–gas reactions. For example, gas–gas reactions can use ammonia (NH_3) to store heat by converting it to nitrogen and hydrogen. The reaction can be faster using different catalysts and under high pressure. Earlier studies investigated SO_3, CH_3OH, C_6H_{12}, and CH_4 for gas–gas reactions; isopropanol and ammonium hydrogen sulfate for liquid–gas reactions; and metal oxides and inorganic oxides (e.g., PbO and MgO) for solid–gas reactions. Figure 5.5 presents the relationship between energy density and the temperature required for charging.

The benefits of chemical-based heat storage technologies are high energy storage density and long-term storage capacity, but they have a high cost due to reactants and complex reaction systems. Also, the reactants can be non-renewable and toxic, posing environmental impacts and safety concerns. Currently, chemical-based

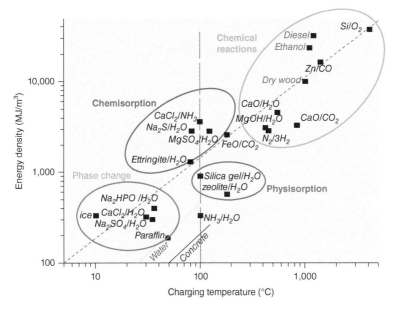

Figure 5.5 Energy density comparison of different materials with charging temperature. Source: Dimitriev et al. [7]/John Wiley & Sons.

heat storage technology has TRL 4 with a small market compared to other heat storage technologies. Further details about chemical-based heat storage and candidate materials (e.g., carbonates, hydroxides, and metal hydrides) have been provided by [3, 11]. Table 5.2 provides a comparison of the most used heat storage technologies.

5.3 Case Studies: Commercial Concentrating Solar Power Plants

Sensible heat storage systems have been used to store over 1 GWh worldwide [3]. Earlier studies compared the energy storage capacity of a combined 100 large-scale battery plants with two commercial CSP plants, i.e., Crescent Dunes and Solana. Their results show that each CSP plant provides more energy storage capacity than the combined 100 large-scale battery plants in the United States at the end of 2017 (Figure 5.6) [3].

Currently, molten salt, concrete, or rocks are the primary media for high-temperature sensible heat storage. For example, the Crescent Dunes CSP plant can generate 110 MW and store thermal energy up to 1.1 GWh using molten nitrate salt. In this CSP plant, molten salt is located in the receiver on top of the

Table 5.2 Comparison of heat storage technologies.

Technology	Mechanism	Materials	Density	Benefits	Drawbacks
Sensible	Storing energy through temperature differences in liquid or solid media	Rock, sand, concrete, and molten salt	200–400 kJ/kg for 200–400 °C (different temperatures)	Higher storage capacity, low-cost media, and high maturity	Lower energy density and higher heat losses
Latent	Storing energy through phase-change materials	Salts, metals, and organics	100–200 kJ/kg for nitrate salts 200–500 kJ/kg for metals 1,000kJ/kg for fluoride salts	Higher energy storage density for low-temperature applications	Potential for corrosion, narrow temperature range, and low maturity
Thermochemical	Storing energy in chemical bonds	Water, clays, and salts	300–6,000 kJ/kg	Lower heat loss and potential for long-term storage applications, compact systems, and larger energy densities	Higher complexity and capital costs, and low maturity

Source: Adapted from Ho and Ambrosini [3].

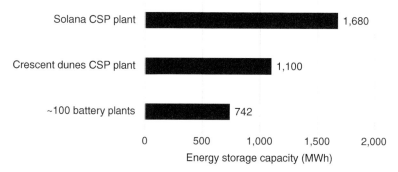

Figure 5.6 Energy storage capacity of combined 100 battery plants with two CSP plants in the United States. Source: Adapted from Ho and Ambrosini [3].

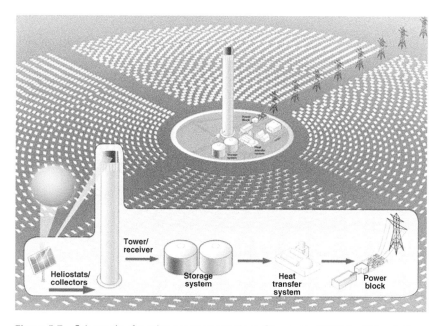

Figure 5.7 Schematic of a solar power tower plant. Source: Ho [12]/John Wiley & Sons.

tower, where it is heated to around 560 °C by concentrated sunlight from a field of heliostats and then flows to hot storage tanks (Figure 5.7). The hot molten salt can later be used in the heat exchanger to heat up water and generate steam to run turbines and generators for power production. This process can be repeated and generate power with large capacity factors.

There are several other CSP plants that use solid media, such as rock, graphite, and concrete. For instance, Siemens Gamesa in Germany developed 130 MW heat storage in 2019, using rocks in a building that can heat up to 600 °C. Graphite Energy in Australia has a 3 MWe heat storage system that uses graphite blocks in the CSP receiver as storage media, which can be used to generate steam and power. Also, EnergyNest in Norway developed a concrete-based heat storage system that can charge and discharge in 30 minutes and store energy for several days with low heat loss, around 2% per day. Some of the challenges of these sensible heat storage systems include: (i) the requirement of a large volume of materials due to their low energy density and (ii) high cost of large capacity storage. Currently, the U.S. DOE capital cost target is $15 per kWh for the entire heat storage system.

Self-Check Questions

1. What are the main types of energy storage systems?
2. What are mature heat storage technologies that have been used on a large scale?
3. What are the main parameters to compare different heat storage technologies?
4. How does sensible heat storage work?
5. What are the benefits and drawbacks of using molten salt for high-temperature heat storage systems?
6. What are the common sensible heat storage media?
7. What are the attributes of sensible heat storage technology?
8. What are the main parameters of sensible heat storage?
9. What is the specific heat capacity?
10. What are the benefits and drawbacks of sensible heat storage technology?
11. How does latent heat storage work?
12. What are the main components of latent heat storage?
13. What are the benefits and drawbacks of latent heat storage technology?
14. What is aqueous salt solution? And how does it work?
15. What are the benefits and drawbacks of the aqueous salt solution for heat storage?
16. How does thermochemical heat storage work?
17. What are the main types and key attributes of thermochemical heat storage?
18. What are the benefits of thermochemical heat storage technologies?
19. How does sorption-based heat storage work?
20. What are the main components and types of sorption-based heat storage?
21. What are the benefits and drawbacks of sorption-based heat storage technology?
22. How does chemical-based heat storage work?
23. What are the main components and factors of chemical-based heat storage technology?

24. What are the main types of chemical-based heat storage technology?
25. What are the benefits and drawbacks of chemical-based heat storage technology?

References

1 Zbair, M. and Bennici, S. (2021). Survey summary on salts hydrates and composites used in thermochemical sorption heat storage: a review. *Energies* 14 (11): 3105.

2 Yang, T., Liu, W., Kramer, G.J., and Sun, Q. (2021). Seasonal thermal energy storage: a techno-economic literature review. *Renewable and Sustainable Energy Reviews* 139: 110732.

3 Ho, C.K. and Ambrosini, A. (2020). *Chapter 12 Thermal Energy Storage Technologies*. Sandia National Laboratories.

4 Olivkar, P.R., Katekar, V.P., Deshmukh, S.S., and Palatkar, S.V. (2022). Effect of sensible heat storage materials on the thermal performance of solar air heaters: state-of-the-art review. *Renewable and Sustainable Energy Reviews* 157: 112085.

5 Nomura, T. and Akiyama, T. (2017). High-temperature latent heat storage technology to utilize exergy of solar heat and industrial exhaust heat. *International Journal of Energy Research* 41 (2): 240–251.

6 Hu, N., Li, Z.-R., Xu, Z.-W., and Fan, L.-W. (2022). Rapid charging for latent heat thermal energy storage: a state-of-the-art review of close-contact melting. *Renewable and Sustainable Energy Reviews* 155: 111918.

7 Dimitriev, O., Yoshida, T., and Sun, H. (2020). Principles of solar energy storage. *Energy Storage* 2 (1): e96.

8 Gbenou, T.R.S., Fopah-Lele, A., and Wang, K. (2022). Macroscopic and microscopic investigations of low-temperature thermochemical heat storage reactors: a review. *Renewable and Sustainable Energy Reviews* 161: 112152.

9 Henninger, S.K., Jeremias, F., Kummer, H., and Janiak, C. (2012, 2012). MOFs for use in adsorption heat pump processes. *European Journal of Inorganic Chemistry* 16: 2625–2634.

10 Yan, T. and Zhang, H. (2022). A critical review of salt hydrates as thermochemical sorption heat storage materials: thermophysical properties and reaction kinetics. *Solar Energy* 242: 157–183.

11 Yan, T., Wang, R.Z., Li, T.X. et al. (2015). A review of promising candidate reactions for chemical heat storage. *Renewable and Sustainable Energy Reviews* 43: 13–31.

12 Ho, C.K. (2014). Computational fluid dynamics for concentrating solar power systems. *WIREs Energy and Environment* 3 (3): 290–300.

6

Biofuel Production

Biofuels are low-emission fuels that can help decarbonize different industrial sectors where electrification is challenging and cannot meet net-zero emission targets. Biofuels can be classified into two main types: gaseous (e.g., biogas and biomethane) and liquid hydrocarbon fuels (e.g., bioethanol and biodiesel). The main challenges are compatibility with the existing end-use technologies (e.g., combustion engines) and distribution infrastructure. Globally, liquid biofuels have been used in road transportation that can be expanded to aviation and shipping. Gaseous biofuels have been used as clean cooking fuels for power and heat generation. Liquid biofuels are mainly produced from crops (e.g., corn, sugarcane, and soybeans) that can compete and interrupt food production, and limit the supply for biofuel production.

Recently, several studies investigated broad feedstocks (e.g., forest or agriculture residues, algae, or wastes) to address the food interruption that can expand biofuel production using different technologies, such as gasification, pyrolysis, and liquefaction processes. Currently, cellulosic ethanol can be used as a drop-in (hydrocarbon fuel) substitute for jet fuel and diesel. Biomass feedstocks are abundant, renewable resources that can be converted to liquid transportation biofuels, power, and thermal energy (Figure 6.1). There are two types of biomass feedstocks: plant-based materials (e.g., crop wastes, forest residues, grasses, woody crops, food wastes, and urban wastes) and algae-based materials (e.g., microalgae and macroalgae) [1]. The existing technologies can reuse carbon from biomass feedstocks to produce low-emission power and bioproducts.

The three main greenhouse gas (GHG) contributors in the agriculture and land use sector for food production are CO_2 from deforestation and land use (9%), CH_4 from livestock (5%) and rice (1%), and N_2O from manure and fertilizers (4%). Recent studies show that the primary deforestation on the Earth is happening in Brazil and Indonesia for producing cattle and soybeans to feed them in Brazil and palm oil for cooking and other purposes in Indonesia. To mitigate GHGs from the agriculture and land use sector, we can reduce food waste (which is a

Net-Zero and Low Carbon Solutions for the Energy Sector: A Guide to Decarbonization Technologies, First Edition. Amin Mirkouei.
© 2024 John Wiley & Sons, Inc. Published 2024 by John Wiley & Sons, Inc.

Biomass to biofuel processes

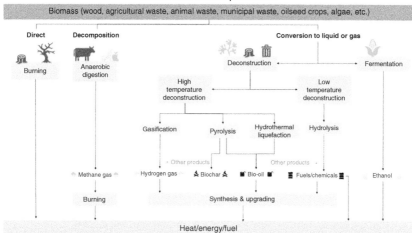

Figure 6.1 The most common biofuel production processes. Source: U.S. Department of Agriculture (public domain) [2].

third of total food production), change diet to more plant-based products, protect ecosystems (forests, grasslands, wetlands, peatlands, oceans, and seas), enhance the efficiency of the food production practices (irrigation, nutrient, and waste management), and produce renewable products (e.g., biofuels and biochemicals). In addition, we can use regenerative annual cropping practices as a net-negative solution to capture CO_2 and put it into the soil or produce renewable energy.

Biofuel production from organic waste streams is a low-emission pathway for energy generation, waste management, and decarbonization through pyrolysis, incineration, or gasification. For example, paper recycling can avoid deforestation, reduce water use, and lower emissions compared to conventional papers made from virgin timbers, as well as consider carbon sequestration from standing trees. Recycled papers can be converted to several products, such as toilet papers, napkins, and newsprints for up to 7 times, depending on the fiber viability. Recycling nonorganic wastes (e.g., glass, plastics, metals, and electronics) can reduce raw materials, energy, and water consumption, and subsequently address released emissions and resource scarcity. For instance, aluminum from recycled materials can reduce around 95% of energy consumption compared to raw materials.

Biofuel production technologies can be mixed with carbon capture, utilization, and sequestration (CCUS) and reduce CO_2 emissions at relatively low capital and operational costs due to the pure CO_2 stream from these processes. According to the latest U.S. DOE studies, over 1 billion tons of nonfood biomass feedstocks are available annually in the United States that can produce up to 50 billion gallons of biofuels, 50 billion pounds of biochemicals and bioproducts, and 85 billion kWh of power for 7 million households, as well as create over a million jobs and promote the U.S. economy [3].

An overview of key low-emission solutions for biofuel production is presented in this chapter. Particularly, this chapter provides 21 technologies with high technology readiness level, along with a case study about gasification with CCUS.

6.1 Biogas

Biogas is a mixture of different gases (e.g., CO_2, CH_4, H_2O, and H_2S) produced from the anaerobic digestion of organic wastes (e.g., food wastes, manure, or sludge) and biomass feedstocks (e.g., forest or agriculture residues) in the absence of oxygen. CH_4 in biogas is relatively high (50–75%) and can be used for heat and power generation. Biogas energy content is similar to that of natural gas since CH_4 is the primary component of natural gas (Table 6.1).

Anaerobic digestion is a chemical process in which microorganisms and bacteria digest (break down) organic wastes in oxygen-free, sealed vessels or reactors (Figure 6.2). This technology can produce biogas and nutrient-rich fertilizers and cut up to 10 gigatons of GHGs. Biogas can be purified by removing other low-value gases (e.g., CO_2 and H_2S) to produce renewable natural gas mixed into natural gas distribution systems, or compressed to generate alternative transportation fuels. The main benefits of biogas production include: (i) generating revenue from waste products, (ii) reducing GHG emissions, and (iii) providing renewable

Table 6.1 Energy content of various fuels.

Fuel	Heating Value (MJ/kg)
Biodiesel	42.2
Butanol	36.6
Diesel	45.4
Dimethyl ether	31.7
Ethanol	29.7
Gasoline	46.5
Hydrogen	141.8
Methanol	22.7
Methane	55.5

Source: Grimalt-Alemany et al. [4]/John Wiley & Sons.

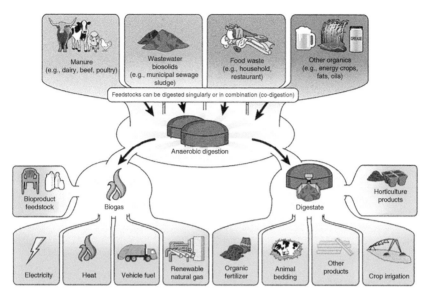

Figure 6.2 A schematic of the anaerobic digestion process and its products. Source: U.S. Environmental Protection Agency [5].

energy sources and byproducts (e.g., fertilizers) from its digestate (leftover materials from the anaerobic digestion process). The major drawbacks are high capital and operational costs and uncertainties about biogas quality due to inconsistency in biomass feedstocks.

Currently, anaerobic digestion technology is the main process pathway for biogas production, and Germany, as the leading producer of biogas, can reduce its reliance on fossil fuels. Large-scale digesters have been used in European countries (e.g., Germany, Italy, the United Kingdom, and Denmark), and small-scale digesters in the United States and Asia (e.g., China). Portable, small digesters in backyards can digest organic (crops and animal) wastes and produce biogas that can be used for cooking instead of charcoal, natural gas, or kerosene, especially in low-income, developing countries in Africa and Asia. Traditional cooking practices, using mainly wood, charcoal, or coal, have several negative environmental impacts, such as forest degradation and GHG emissions. Efficient cooking stoves that use renewable energy sources can cut over 70 gigatons of GHG emissions by 2050.

6.1.1 Anaerobic Digestion of Non-Algae Feedstocks (TRL 9)

Agricultural residues, food and organic wastes, and sewage sludge are the main types of non-algae feedstocks that can be converted into organic acids and then

Figure 6.3 Biogas production on a farm processing cow dung. Source: Wiley Digital Library.

a mixture of different gases during the anaerobic digestion process (Figure 6.3). This process includes several stages, such as hydrolysis (large molecules to small ones), acidogenesis (small molecules to organic acids), acetogenesis (organic acids to acetate), and methanogenesis (acetate to CH_4 and CO_2). Small molecules (e.g., sugars, proteins, and fats) can produce organic acids, such as amino acids, and short-chain fatty acids (e.g., acetic acids). The main parameters that can impact the process performance are organisms type, loading rate, retention time, pH, and temperature. The primary benefit of this process is reusing organic waste instead of sending them to landfills, but some of the drawbacks are high capital and operational costs, different feedstock composition, and feedstock availability throughout the year.

Non-algae feedstock handling and treatment through anaerobic digestion can address several socio-environmental challenges, mainly waste generation from anthropogenic activities. Several studies investigated biogas production from organic wastes (e.g., food wastes and sludge) that can be used for various purposes, such as power, heat, or biomethane (CH_4) production. Globally, China, India, and the United States are the top three food waste and sludge generators [6]. The liquid and solid materials from an anaerobic digester can be used for producing advanced biochemicals, such as bioplastics, animal bedding, and soil conditioners. This solution has several benefits for farmers, such as waste management, emission reduction, diversified revenues, land conservation, and rural economic growth. Some of the leading producers of anaerobic digestion for biogas production from non-algae feedstocks are EnviTec Biogas, ENGINE, and PlanET Biogas.

6.1.2 Anaerobic Digestion of Algae (TRL 4)

Anaerobic digestion technology has been used to process or break down algae by microorganisms and bacteria without oxygen. The main product of this process is biogas, including CO_2 and CH_4 that can be used for renewable energy production. The anaerobic digestion process of algae is similar to that of non-algae feedstocks, involving four steps: hydrolysis, acidogenesis, acetogenesis, and methanogenesis. During this process, algae (complex organic compounds) are converted to simple compounds, such as carbohydrates, proteins, or lipids, by bacteria or hydrolytic enzymes through hydrolysis. Then acidogenic bacteria convert those simple compounds to short-chain fatty acids (e.g., acetic or butyric acids), and then acetogenic bacteria convert fatty acids to acetate, hydrogen, and CO_2 that can be converted to methane by methanogenic archaea.

The key parameters impacting the process yield are temperature, pH, algae loading rate, and process time. In this process, the temperature range is 25–40 °C, pH range is 6.5–8.5, and the process time is around 20–30 days. The main challenge of algae anaerobic digestion is the inconsistency in algal components and compositions (e.g., lignin type, cellulosic fibers, and polyphenols) that can limit digestibility and reduce process yield. Additionally, seasonal growth and location are other associated problems. This technology is still in the early stages of development and has TRL 4. Some of the companies that have been involved in the anaerobic digestion of algae for renewable energy production (e.g., biogas) are Algenol Biofuels, ENN Group, Algix, and Heliae Development. More detailed information about the anaerobic digestion of algae has been provided by [7].

6.2 Biomethane

Methane-based energy from biomass (biomethane) is a low-emission solution that captures emissions and generates energy from organic wastes in landfills instead of using fossil fuels (e.g., coal and natural gas). Landfills release over 10% of methane worldwide due to decomposing organic wastes, such as food, wood, or paper. Methane has 34 times more warming effects than CO_2 over 100 years.

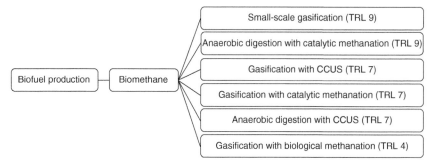

Figure 6.4 Classification of the most used biomethane production technologies.

Biomethane production is mostly achieved through the upgrading of biogas (removing CO_2 and other gases) to pure methane, or gasification and anaerobic digestion of various feedstocks, such as municipal solid wastes, manure, or agriculture residues (Figure 6.4) [8, 9].

Gasification is a deconstruction (endothermic) process under high temperatures (over 700 °C) with low or partial oxygen present. Figure 6.5 presents gasification and upgrading processes for power, fuel, and chemical production from various feedstocks (e.g., biomass, coal, and organic wastes). The main product of gasification is synthetic gas (syngas), consisting of hydrogen and CO. Gasifiers have been used for centuries and can be classified into three main types

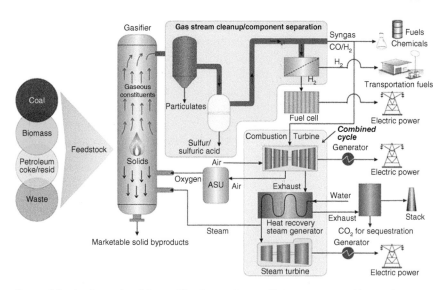

Figure 6.5 A schematic of the gasification and upgrading processes and its products. Source: Adapted from NETL [10].

based on the materials' contacting method with gasification reagents (heat and air): moving-bed (e.g., updraft or downdraft), fluidized-bed, and entrained-bed reactors. The key parameters for evaluating the gasification process are operating conditions (e.g., temperature and pressure), process capacity (e.g., feed rate and particle size), syngas composition before and after cleaning, power output, total products throughput, emissions, water consumption in the cooling tower, and contaminants in wastewater.

Gasification syngas (a mixture of CO and H_2) can be used for the following: (i) power generation, using gas turbines and integrated coal gasification combined cycle or steam power and Rankine cycle; (ii) hydrogen generation, using fuel cells; (iii) production of methanol or similar chemicals, such as acetic acid, polyolefins, methyl esters, and formaldehyde; (iv) production of Fischer–Tropsch (FT) process products, such as naphtha, wax, diesel, and gasoline, and (v) production of other products, such as ethanol and synthetic natural gas (SNG). However, combustion is an exothermic process that needs high oxygen to produce high-temperature gas or power and subsequently release large amounts of emissions [10]. More detailed information about the gasification process has been provided by [10].

6.2.1 Small-Scale Gasification (TRL 9)

Gasification process involves four steps: (i) drying (reducing the moisture content in the materials), (ii) pyrolysis (breaking down materials into small molecules), (iii) combustion (producing heat and gases in partial oxygen present), and (iv) reduction (removing CO_2 and other impurities using reduction agent). The outputs of the pyrolysis process are CO, H_2, and CH_4. The key factors are feedstock type and properties due to their chemical compositions, process temperature, reduction agent for producing clean syngas, and residence time. These factors can affect reaction kinetics, process yields (conversion efficiencies), and product quality (e.g., gas composition and heating value).

Small-scale gasification plants have been growing and operating, especially in the European market, mostly in Italy and Germany. The main reactor types used for small-scale processes are downdraft and updraft gasifiers. Small-scale gasification has been commercialized and has high TRL due to the high efficiency that can convert a wide range of feedstocks to bioenergy (e.g., biomethane) and reduce GHG emissions and dependency on fossil fuels. As a result, small-scale gasification can be cost-effective and less complex compared to large-scale processes. Some leading producers of small-scale gasification plants are EQTEC, Cortus Energy, Waste to Energy International, Sierra Energy, and Nexterra. Patuzzi et al. conducted a literature review on small-scale biomass gasification systems [11].

6.2.2 Anaerobic Digestion and Catalytic Methanation with Hydrogen (TRL 9)

Anaerobic digestion is a multistep process in which microorganisms break down organic materials in the absence of oxygen. The key parameters in the anaerobic digestion process are organic loading rate, temperature, pH, volatile fatty acid, residence time, and carbon/nitrogen ratio. The dominant organic waste management and recycling methods are landfill, incineration, composting, and anaerobic digestion. Composting and anaerobic digestion require pretreatment processes. Organic waste recycling can be used for power, biogas (e.g., methane and CO_2), and composting generation [12, 13]. As explained earlier in biogas production (Section 6.1), there are different pathways for methane and carbon dioxide production from organic wastes (e.g., food) through anaerobic digestion. The main steps include hydrolysis (to produce glucose, glycerol, and amino acids), acidogenesis (to produce volatile fatty acids), acetogenesis (to produce acetate, hydrogen, and carbon dioxide), and methanogenesis (to produce CH_4 and CO_2) [9, 12].

Catalytic methanation with hydrogen after anaerobic digestion can produce biomethane from different organic materials. Particularly, the anaerobic digestion process converts organic materials to CH_4, CO_2, H_2O, and NH_3. Then CO_2 from the anaerobic digestion process is reacted with hydrogen using a catalyst during catalytic methanation with hydrogen ($CO_2 + 4H_2 \rightarrow CH_4 + H_2O$) and produces methane and water. The main benefits are bioenergy production to reduce dependency on fossil fuels, waste reduction, and GHG emissions reduction, especially capturing and utilizing methane. The drawbacks are high capital costs, high energy requirements for hydrogen production, and process complexity. Integrated anaerobic digestion and catalytic methanation with hydrogen technology is commercialized and has TRL 9. Some leading producers of this technology for biomethane production in the United States are Dominion Energy, DMT Clear Gas Solutions, and Brightmark.

6.2.3 Gasification with CCUS (TRL 7)

Gasification process can convert various feedstocks and wastes to syngas using gasifying agents (e.g., oxygen or steam) at high temperatures. Then syngas can be purified for biomethane production that can be used as a renewable fuel for transportation or power generation. The chemical reactions during the gasification process of organic materials (e.g., woody biomass or agriculture waste) are as follows:

$$C_6H_{12}O_6 + 6O_2 \rightarrow 6CO_2 + 6H_2O \tag{6.1}$$

$$C_6H_{12}O_6 \rightarrow 3CH_4 + 3CO_2 \tag{6.2}$$

Then the methane and CO_2 can be separated, using various carbon-capturing methods, such as absorption, pressure swing adsorption, or membrane separation. As explained in Chapter 2, CCUS is a process that can capture CO_2 from various sources (e.g., coal, biomass, and natural gas) and conversion processes (e.g., gasification), reuse it for other applications or processes (e.g., chemical production or enhanced oil recovery), and store it before entering into the atmosphere. Several studies investigated different processes (e.g., biomass and waste gasification) with catalyst methods (e.g., thermo-, plasma-, or photo-catalytic) for improving CO_2 methanation processes because synthesized methane is compatible with natural gas and can be distributed using the existing pipelines and infrastructures [14]. Ni-based catalysts are the most widely used for CO_2 methanation due to their low cost and high efficiency.

Overall, gasification with CCUS can be used for biomethane production from various sources. The main benefits are bioenergy production, GHG emission (e.g., CO_2) mitigation, and waste reduction. The captured CO_2 from the CCUS process can be used in various applications (e.g., enhanced oil recovery). The drawbacks are process complexity and uncertainties due to feedstock variability. This technology has TRL 7 and requires further improvement to be fully commercialized.

6.2.4 Gasification with Catalytic Methanation (TRL 7)

Syngas from the gasification process contains CO, H_2, CH_4, and CO_2. The catalytic methanation process can convert CO and H_2 to methane and water ($CO + 3H_2 \rightarrow CH_4 + H_2O$), which can increase the methane content of syngas. The chemical reaction formula for gasification and catalytic methanation from biomass is as follows:

$$C_6H_{12}O_6 + 6O_2 + 3CO + 9H_2 \rightarrow 3CH_4 + 6CO_2 + 9H_2O$$

Gasification with catalytic methanation conversion process can use different methods (e.g., air separation during the gasification process) to correct the carbon/hydrogen ratio in syngas and increase methane purity, which is a highly exothermic process (loss of energy, especially heat). Currently, selective catalytic methanation of syngas from gasification has been commercialized using different catalysts, such as nickel. However, the economic viability highly depends on the availability of natural gas and its price. Some countries (e.g., China and India) with high natural gas prices can use this pathway [10]. Overall, gasification with catalytic methanation conversion pathway can produce biomethane from different feedstocks, and its main product (syngas) can be used as renewable natural gas. Similar to other biomethane production processes, this process has several benefits, such as bioenergy production and waste management, but the main drawbacks are high cost and process complexity.

6.2.5 Anaerobic Digestion with CCUS (TRL 7)

As explained earlier, CCUS is a decarbonization strategy that can be used for power generation from coal, natural gas, and biomass, as well as biofuel production, such as biomethane. One of the potential conversion pathways is the biomethanation of CO_2 from the anaerobic digestion process ($C_6H_{12}O_6 \rightarrow 3CH_4 + 3CO_2 + 3H_2O$), which depends on the feedstock type (food waste, livestock manure, or sludge). The methanation process of CO_2 consists of the microbial transformation of CO_2 from organic materials, along with hydrogen (H_2) to methane (CH_4) and water (H_2O). The benefits and drawbacks of this process are similar to those of other biomethane production processes, such as gasification with CCUS.

6.2.6 Gasification with Biological Methanation (TRL 4)

Gasification with biological methanation pathway combines the benefits of both thermochemical and chemical conversion processes for renewable biomethane production from waste streams. Biological methanation of syngas from gasification is an anaerobic process in which bacteria break down syngas into methane and CO_2 using various microbial groups, such as methanogenic archaea. The biological methanation process has several benefits compared to the catalytic methanation process, such as milder operation conditions (e.g., low temperature and pressure) and less sensitivity to syngas purity and carbon/hydrogen ratio.

The chemical reactions during biomethane production pathways from syngas (CO, CO_2, and H_2) are as follows:

- Water–gas shift (WGS) reaction ($CO + H_2O \rightarrow CO_2 + H_2$) and hydrogenotrophic methanation ($HCO_3^- + 4H_2 + H^+ \rightarrow CH_4 + 3H_2O$)
- Direct hydrogenotrophic methanation from syngas (CO_2/H_2)
- WGS and homoacetogenesis ($2HCO_3^- + 4H_2 + H^+ \rightarrow CH_3COO^- + 4H_2O$)
- Homoacetogenesis and acetoclastic methanation ($CH_3COO^- + H_2O \rightarrow CH_4 + HCO_3^-$)
- Fatty-acid production, fatty-acid oxidation, and acetoclastic methanation
- Carboxydotrophic acetogenesis ($4CO + 4H_2O \rightarrow CH_2COO^- + 2HCO_3 + 3H^+$) and acetoclastic methanation
- Direct carboxydotrophic methanation ($4CO + 2H_2O \rightarrow 3CO_2 + CH_4$)

Gasification with biological methanation conversion pathway has not been commercialized yet, but it can be economically feasible for small-scale gasification plants due to lower operational costs. The drawbacks for reaching the commercial stage are methane productivity and conversion efficiency due to the high moisture content of waste streams. Further details about the biological methanation of syngas have been provided by [4].

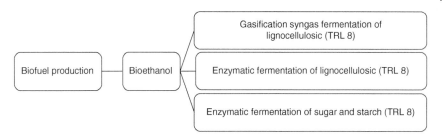

Figure 6.6   Classification of mature bioethanol production technologies.

6.3   Bioethanol

Bioethanol (CH_3CH_2OH) is a renewable hydrocarbon fuel from biomass feedstocks that can be mixed with fossil fuels (e.g., gasoline and diesel) to increase their octane and reduce CO_2 emissions by oxygenating the fuel [15]. The production method can be classified into two main processes, i.e., chemical and thermochemical, which depend on the feedstock type (Figure 6.6). The sugar- or starch-based feedstocks have a shorter process than lignocellulosic feedstocks. In the United States, most bioethanol is produced from plant starches and sugars (e.g., corn starch) using fermentation, owing to a positive energy balance. In other words, less energy is required for bioethanol production than its energy content. However, enormous efforts have been invested in lignocellulosic feedstocks (e.g., wood residues) and wastes due to water and fertilizer requirements for corn production and lower emission levels [16]. The most common blend (known as E10) contains 10% bioethanol and 90% gasoline that can be used in the existing internal combustion engines, but it highly depends on the season and geography. Melendez et al. provided more details about bioethanol production methods and technologies [17].

6.3.1   Gasification Syngas Fermentation of Lignocellulosic Feedstocks (TRL 8)

Bioethanol production from lignocellulosic feedstocks (e.g., woody biomass) through the gasification process involves a high-temperature process to produce syngas, and fermentation to upgrade syngas to bioethanol and other byproducts (Figure 6.7) [19]. Particularly, it involves three main steps: (i) biomass gasification, (ii) syngas fermentation, and (iii) bioethanol distillation. As explained earlier in biomethane production, air, oxygen, or steam can be used as a gasifying agent. Then syngas from gasification is fed to the fermenter to convert CO and H_2 to bioethanol and other byproducts via microorganisms (e.g., yeast or bacteria). After fermentation, bioethanol can be purified through a distillation process,

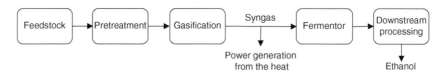

Figure 6.7 Process flows of bioethanol production through gasification syngas fermentation of lignocellulosic feedstocks. Source: Adapted from Kennes et al. [18].

including heating and condensing steps. The overall chemical reactions of gasification syngas fermentation of lignocellulosic feedstocks are as follows:

- Gasification: lignocellulosic biomass $(C_6H_{10}O_5) + 2H_2O + 2O_2 \rightarrow 6CO + 10H_2$
- Syngas fermentation: $4H_2 + 2CO \rightarrow CH_3CH_2OH + 2H_2O$

Overall, gasification syngas fermentation of lignocellulosic feedstocks for bioethanol production has several benefits, such as bioenergy production from abundant and renewable sources (e.g., biomass feedstocks) and lower GHG emissions and environmental impacts. However, this production pathway requires complex chemical and physical processes, high capital and operational costs, and high land use. Also, bioethanol has lower energy density compared to gasoline, leading to lower efficiency, higher fuel consumption, and higher costs. More detailed information about syngas fermentation has been provided by NETL [10].

6.3.2 Enzymatic Fermentation of Lignocellulosic Feedstocks (TRL 8)

Enzymatic fermentation is a common method for converting lignocellulosic feedstocks (e.g., woody biomass) into bioethanol through microorganisms (e.g., yeast and bacteria) that can metabolize complex sugars, break them down to simpler sugars, and produce bioethanol [17]. The chemical process includes pretreatment and hydrolysis to release and break down sugars, which are fermented by microorganisms to bioethanol and lignin (Figure 6.8). Particularly, enzymes help breaking down the hemicellulose and cellulose in lignocellulosic feedstocks to sugars during hydrolysis.

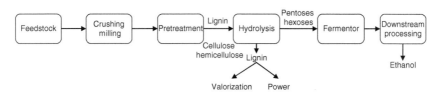

Figure 6.8 Process flows of bioethanol production through enzymatic fermentation of lignocellulosic feedstocks. Source: Adapted from Kennes et al. [18].

The chemical reactions of enzymatic fermentation of lignocellulosic feedstocks are as follows:

- Hydrolysis: Cellulose + water → glucose and hemicellulose + water → xylose, arabinose, galactose, and simple sugars
- Fermentation: Glucose (or simple sugars) → $2CH_3CH_2OH + 2CO_2$

This production pathway can reach up to 80% process yields. The drawbacks are pretreatment requirements, high water use for hydrolysis, and high processing costs due to enzyme requirements. Also, bioethanol production from lignocellulosic feedstocks (e.g., wood and grass) requires more processes than from starch-based crops. Melendez et al. provided more details about the fermentation method for bioethanol production [17].

6.3.3 Enzymatic Fermentation of Sugar and Starch (TRL 8)

Enzymatic fermentation of sugar- and starch-based crops (e.g., corn starch) can produce bioethanol by either dry or wet milling. Dry milling is mostly used due to lower capital costs, including grinding the feedstocks to powder and fermenting them to ethanol, CO_2, and byproducts. Wet milling needs more capital cost and includes separating starch, fiber, and protein before processing and producing bioethanol and other byproducts. This process uses enzymes (e.g., alpha-amylase and glucoamylase) to convert complex carbohydrates or disaccharides from starch or sucrose, respectively, into simple sugars (e.g., maltose, fructose, and glucose) and then fermented by yeast to produce bioethanol. The chemical reaction of fermentation of glucose is as follows:

$$C_6H_{12}O_6 \rightarrow 2C_2H_5OH + 2CO_2$$

The benefits of enzymatic fermentation for bioethanol production include: (i) higher efficiency than conventional fermentation methods due to the use of enzymes, (ii) lower energy consumption and temperatures, (iii) higher process yields compared to conventional fermentation methods, (iv) lower environmental impacts and emissions, and 5) lower processing costs due to higher yield and efficiency. The drawbacks are: (1) high enzymes cost, (ii) process complexity due to enzyme sensitivity to temperature and pH changes, and (iii) longer production time that can reduce the capacity. Scaling up the enzymatic fermentation process can be challenging due to the process complexity and several parameters. Currently, this technology has TRL 8, and new enzymes and processes can increase its maturity. Some leading producers of the enzymatic fermentation process are Novozymes, Abengoa Bioenergy, and DuPont Industrial Biosciences.

6.4 Biodiesel

Biodiesel is a nontoxic, biodegradable liquid biofuel, produced from renewable sources, such as animal fats, vegetable oils (e.g., soybean corn and cotton seed oil),

Table 6.2 Physical and chemical characteristics comparison of biodiesel and diesel #2.

Physical characteristics	Biodiesel	Diesel #2	Unit
Carbon	77	87	w%
Hydrogen	12	13	w%
Oxygen	11	0	w%
Sulfur	15	0–15	ppm
Density	7.3	7.1	Ib/gal
Boiling point	315–350	180–340	°C
Flash point	100–170	60–80	°C
Higher heating value	127,960	138,490	Btu/gal
Viscosity	4–6	1–4	Mm^2/s
Cetane number	47–65	40–55	–

and recycled cooking oils. Biodiesel is similar to petroleum-based diesel that can be blended with diesel and used in existing engines (Table 6.2). The most common blend (known as B20) contains 20% biodiesel and 80% petroleum diesel. Biodiesel production is based on converting fats and oils to long-chain mono alkyl esters or fatty acid methyl esters (FAME).

The cetane number indicates the ignition quality of the fuel (e.g., biodiesel or diesel) and how well it ignites in the engine. Higher quality fuel has a higher cetane number that has more efficiency, lower emissions, and better performance. The benefits of biodiesel compared to other biofuels include ease of use, low GHG emissions, stability in extended storage, and improved engine operation due to increased cetane number. The drawbacks are higher production costs compared to diesel, limited feedstocks, compatibility issues with existing engines and infrastructure, and environmental impacts due to high land use and resources.

There are several methods for biodiesel production from different feedstocks, and the main methods use gasification, pyrolysis, transesterification, or liquefaction technologies (Figure 6.9). More detailed information about biodiesel production has been provided by [20].

6.4.1 Esterification of Fatty Acids (TRL 10)

Esterification is one of the conversion processes for fatty acid methyl ester or biodiesel production by converting free fatty (carboxylic) acids in vegetable oils or animal fats into methyl esters in the presence of catalysts (Figure 6.10). This process uses short-chain alcohols (e.g., methanol) and catalysts (e.g., potassium hydroxide or sodium hydroxide) to convert oil to biodiesel and glycerin, which is a co-product of the esterification process. The key parameters in this process are water content, reaction temperature and time, and alcohol-to-acid mixing ratio.

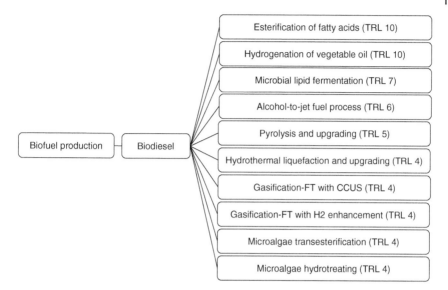

Figure 6.9 Classification of mature biodiesel production technologies.

Reducing water content can increase the esterification reaction and conversion rates. The reaction temperature is around 55–60 °C, and it takes several hours to complete the esterification process. The longer process can increase the process yield; however, very long processes can lead to side reactions that form other byproducts.

The key components of the esterification process are feedstock, alcohol, catalysts, conversion reactors, separators, and distillation columns. This process involves two main chemical reactions, which are as follows:

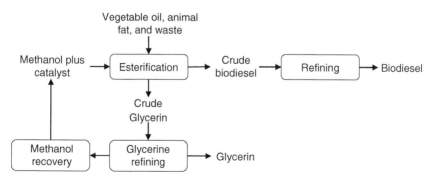

Figure 6.10 Process flow of basic esterification process. Source: Alleman et al. [20].

- Fatty acid esterification: fatty acids $+ CH_3OH \rightarrow$ fatty acid methyl ester $+ H_2O$
- Triglyceride transesterification: triglycerides $+ 3CH_3OH \rightarrow 3$ fatty acid methyl ester $+$ glycerol

The esterification process can be optimized by high temperatures and longer reaction times, as well as using excess methanol. Also, transesterification can be optimized similarly to the esterification process parameters, except for a shorter reaction time. The main benefit of this process is low GHG emissions, using domestic renewable feedstocks (e.g., cooking oil, animal fats, and vegetable oils) compared to fossil-based diesel. The drawbacks are high production costs, land-use change, and competition with food production. Biodiesel producers (e.g., Louis Dreyfus Company and Archer Daniels Midland Company) use various feedstocks (e.g., palm oil and soybean oil) and can be used in different sectors, such as agriculture, transportation, and industry.

6.4.2 Hydrogenation of Vegetable Oil (TRL 10)

Hydrogenated vegetable oil is a paraffinic fuel without sulfur and aromatics, and its production process includes the hydrogenation of animal fats and vegetable oils with lower energy density and higher cetane number than petroleum-derived diesel fuel. Earlier studies show that hydrogenation of vegetable oil is one of the feasible solutions for advanced biofuel production, consisting of various processes, such as triglyceride hydrogenation to decompose vegetable oil to monoglycerides, diglycerides, and carboxylic acids, and then convert them into alkanes via decarboxylation and hydrogenation at high pressure (over 400 psi) and temperature around 350 °C using various catalysts for hydroprocessing.

Generally, hydrogenation uses hydrogen gas to convert unsaturated fatty acids to saturated fatty acid chains. This process can improve the properties of fatty acid methyl esters and their stability using hydrogen gas or donor in the presence of a catalyst (e.g., palladium or nickel). The key parameters of the hydrogenation process are reaction temperature, pressure, and catalyst type. Improving biodiesel properties makes it a suitable alternative in cold regions and reduces degradation risks during storage. The main drawbacks are the production costs and the environmental impacts of trans fatty acids, such as human health effects.

Hydrogenated vegetable oil has received considerable attention recently due to its application as a drop-in fuel in the existing combustion engines as well as lower emissions than diesel (up to 50%). Additionally, hydrogenated vegetable oil has higher oxidation stability, energy content, and heat release rate due to higher cetane number than fatty acid methyl esters. Some leading producers are Renewable Energy Group, Archer Daniels Midland Company, and Louis Dreyfus Company. Several countries (e.g., Japan, Finland, the United Kingdom, Italy, the

United States, and Korea) have developed hydrogenated vegetable oil production. More detailed information can be found in [21].

6.4.3 Microbial Lipid Fermentation (TRL 7)

The microbial lipid fermentation process uses microorganisms (e.g., algae, yeast, or bacteria) to convert sugars (e.g., glucose or sucrose) or other carbon sources into lipids and hydrocarbons that can be upgraded through a complex chemical (metabolic) reaction (e.g., hydroprocessing) to synthetic iso-paraffin or biodiesel. Generally, lipids are made of glycerol and fatty acid chains, consisting of a hydro-carbon chain with a carboxylic acid group. The critical parameters are conversion process configuration (e.g., temperature and pH), microorganism type, and sugar type.

The benefits of this process include: (i) high conversion efficiency for using microorganisms to produce synthetic iso-paraffin from a wide range of carbon sources (e.g., plant, food, and agriculture residues), (ii) lower GHG emissions due to the use of renewable carbon sources compared to traditional products with similar characteristics, and (iii) variety of applications, such as fuel additive to improve diesel performance, solvent for industrial adhesives, and lubricant for high-performance engines. The drawbacks are high production costs and scale-up challenges. Some of the leading producers of microbial lipid fermentation processes for synthetic iso-paraffin production are TerraVia, DSM, and Evolva.

6.4.4 Alcohol-to-Jet Fuel Process (TRL 6)

Alcohol-to-jet fuel (aviation biofuel) process involves the conversion of biomass feedstocks (e.g., lignocellulosic materials or organic matter derived from plants and animals) through biochemical processes into a longer chain of hydrocarbons and alcohols. The biochemical processes use microorganisms (e.g., bacteria or yeast) to break down the biomass into fatty acids and alcohols. Then the alcohols can be further processed through dehydration, oligomerization, and hydrogenation processes to produce jet fuel or biodiesel (Figure 6.11).

Generally, the conversion process includes fermentation and upgrading, and the key parameters are biomass and microorganism types, and reactor design. The benefits of alcohol-to-jet fuel processes are similar to microbial lipid formation, such as lower emission, higher yield, and improved waste management. However, the main drawbacks are complex chemical reactions, process efficiency and optimization challenges, and total costs compared to other biofuels. Several companies (e.g., LanzaTech and Velocys) are trying to improve the alcohol-to-jet fuel production process and commercialize it to use in various sectors, such as the aviation industry. More information about alcohol-to-jet fuel processes can be found in [22].

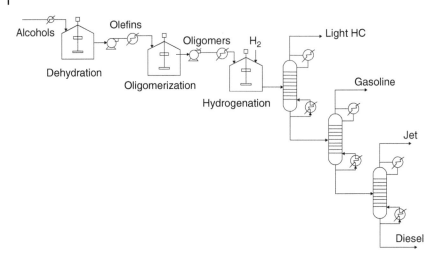

Figure 6.11 Alcohol-to-jet fuel pathway. Source: Wang et al. [22].

6.4.5 Pyrolysis and Upgrading (TRL 5)

Pyrolysis is a thermochemical process (similar to gasification) that can break down organic materials rapidly (in a few seconds), such as biomass feedstocks, to pyrolysis oil, pyrolysis char, and pyrolysis gas under high temperatures (400–700 °C) in the absence of oxygen. Pyrolysis oil can be upgraded using different technologies (e.g., deoxygenation, hydrogenation, or catalytic cracking) to biofuels, e.g., biodiesel (Figure 6.12).

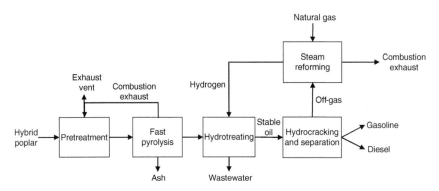

Figure 6.12 Block diagram of fast pyrolysis and upgrading processes. Source: Jones et al. [23].

The chemical reactions during the conversion of biomass feedstocks are complex and depend on biomass type, process conditions (e.g., temperature and pressure), reactor design (e.g., free fall or fluidized bed), and catalyst types. Similar to other biofuel production technologies, the main benefits are fuel production from renewable resources, lower carbon emissions, and effective waste recycling and management. The drawbacks are high capital costs, competition with other biofuels, and complex chemical reactions for different biomass types. Several companies (e.g., Ensyn and BTG Bioliquids) use pyrolysis and upgrading processes for biofuel production. However, these conversion processes still need further improvement to compete with other biofuels, such as bioethanol, in terms of cost and process optimization. Further details about biofuel production via the pyrolysis process have been provided by [24].

6.4.6 Hydrothermal Liquefaction and Upgrading (TRL 4)

Hydrothermal liquefaction is a thermochemical process (similar to pyrolysis) that can deconstruct high moisture content (wet) feedstocks (e.g., algae) to bio-oil under high pressure and moderate temperature ($200–400\,°C$). Similar to other thermochemical processes, the chemical reactions depend on the biomass type, process configurations, and catalyst type. The main product of hydrothermal liquefaction is bio-oil that can be upgraded using different processes, such as hydrodeoxygenation or catalytic cracking with catalysts for removing unwanted compounds and producing compatible biofuels (Figure 6.13).

The benefits and drawbacks are very similar to other thermochemical conversion pathways, such as lower environmental impacts compared to fossil-based transportation fuels, but higher production costs and complex conversion processes, which make it difficult to compete with other fuels. This conversion pathway for biodiesel production has TRL 4, which requires further investigation and

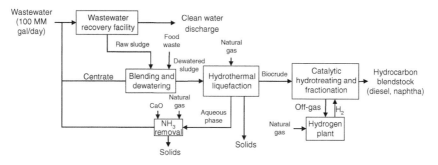

Figure 6.13 Block diagram of hydrothermal liquefaction and upgrading processes. Source: Adapted from Snowden-swan et al. [25].

improvement to meet the market needs. More detailed information about biofuel production through hydrothermal liquefaction and upgrading can be found in [26, 27].

6.4.7 Gasification and Fischer–Tropsch with CCUS (TRL 4)

As explained earlier in biomethane production, the gasification process is a thermochemical process that can convert various biomass feedstocks into syngas at high temperatures. Then syngas can be purified for biodiesel production. During gasification, biomass feedstock is converted to syngas (a mixture of CO, H_2, and other gases). Then FT process can be used to convert syngas into liquid hydrocarbons (e.g., biodiesel). As explained earlier, CCUS is a process that can capture CO_2 from various sources (e.g., biomass, natural gas, and coal), and reuse it for other applications or processes (e.g., enhanced oil recovery) or store it before entering the atmosphere. Several studies investigated different processes (e.g., gasification) with catalysts (e.g., nickel) for improving CO_2-capturing processes [14].

To combine gasification-FT with CCUS, the syngas and CO_2 can be used as feedstocks for liquid hydrocarbon production. The overall chemical reactions are as follows:

- Gasification-FT without CCUS: $nCO + nH_2 \rightarrow C_nH_n + nH_2O$
- Gasification-FT with CCUS: $nCO + nH_2 + nCO_2 \rightarrow C_nH_nCOOH + nH_2O$

Gasification-FT mixed with CCUS can reduce the environmental impacts of biodiesel production from biomass feedstocks. The main benefits are bioenergy production, GHG emission (e.g., CO_2) mitigation, and waste reduction. The captured CO_2 from the CCUS process can be used in various applications (e.g., enhanced oil recovery). The drawbacks are process complexity and uncertainties due to feedstock variability. This conversion pathway has TRL 4 and requires further improvement to be fully commercialized. Further innovations are needed to reduce emissions and costs, and improve the process yield. More detailed information about gasification-FT with CCUS has been provided by NETL [10].

6.4.8 Gasification and Fischer–Tropsch with Hydrogen Enhancement (TRL 4)

Gasification-FT mixed with hydrogen enhancement is a thermochemical pathway for biodiesel production, using several processes, such as gasification, syngas conditioning, FT, and hydrotreatment (Figure 6.14). However, this conversion pathway is not commercially viable due to several barriers and uncertainties regarding biomass type and their properties.

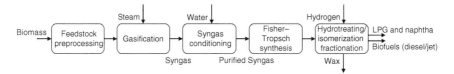

Figure 6.14 Block diagram of gasification and FT with hydrotreating processes for biofuel production. Source: Wang et al. [22].

As explained earlier, gasification can convert organic materials to syngas at high temperatures. Syngas is a mixture of several gases, such as CO and H_2. Then FT process can convert syngas to liquid hydrocarbon fuels, and the hydrogen enhancement process can improve the quality of the hydrocarbon fuels, using additional hydrogen. Particularly, hydrogen enhancement uses different methods, such as hydrodeoxygenation, to remove oxygen and increase hydrogen to produce a fuel similar to fossil-based fuels (e.g., diesel). The chemical reaction of hydrodeoxygenation is as follows:

$$C_nH_nO + nH_2 \rightarrow C_nH_n + nH_2O$$

Overall, gasification-FT mixed with hydrogen enhancement can produce renewable hydrocarbon fuels from biomass feedstocks. Similar to other thermochemical conversion pathways, the main benefits are bioenergy production from renewable resources, GHG emission mitigation, and waste management. The drawbacks are process complexity and uncertainties due to feedstock variability. This conversion pathway has TRL 4 and requires further improvement to be fully commercialized. Further research and developments are needed to improve the process yield and reduce the total costs. The latest study by NETL provided more information about gasification-FT mixed with hydrogen enhancement [10].

6.4.9 Microalgae Transesterification (TRL 4)

Microalgae are photosynthetic organisms with high lipid content that can be converted into biodiesel, using transesterification. Transesterification is a low-temperature chemical process that uses alcohol in the presence of a catalyst to open up the physical structure of feedstocks (e.g., microalgae) and make sugar polymers (e.g., hemicellulose and cellulose) available for producing biodiesel or fatty acid methyl esters and byproducts, such as glycerin [28]. The general chemical reaction is as follows:

- Microalgal lipids (triglyceride) + alcohol → biodiesel + glycerol

This conversion process includes several steps: lipid extraction and purification, catalysis, and biodiesel separation and purification from glycerol (Figure 6.15).

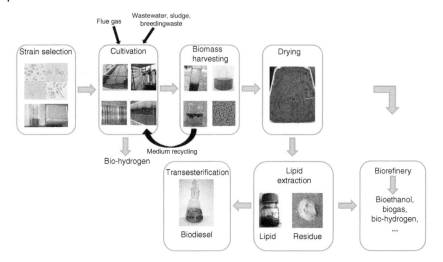

Figure 6.15 Process flow of microalgae transesterification for biodiesel production. Source: Chen et al. [29]/Wiley Online Library.

During catalysis, base catalysts (e.g., sodium hydroxide or potassium hydroxide) are mainly used in transesterification due to higher process yield.

Overall, microalgae transesterification can convert lipids in microalgae into biodiesel, mainly in the presence of a base catalyst. The main parameters are lipid content in microalgae, alcohol and catalyst type, and process configuration (e.g., temperature and time). The benefits are high process yield and the potential to recycle algae. The drawbacks are energy-intensive processes, environmental impacts of large-scale algae cultivation, scale-up challenges, and high capital and operational costs. Some environmental impacts are high nutrient and water use, and GHG emissions from the energy-intensive processes. This technology is still in the early development stages, and several companies are trying to commercialize it for biodiesel production. More detailed information about the microalgae transesterification conversion process can be found in [30].

6.4.10 Microalgae Hydrotreating (TRL 4)

Hydrotreating is a hydrogenation process that can convert high-moisture-content feedstocks (e.g., microalgae) under high temperature and pressure in the presence of catalysts (Figure 6.16). In this process, triglycerides present in microalgae lipids can be converted into hydrocarbons, and the overall chemical reaction is as follows:

- Triglyceride $+ n\mathrm{H}_2 \rightarrow$ hydrocarbon $+ n\mathrm{H}_2\mathrm{O}$

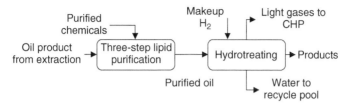

Figure 6.16 Process flow of microalgae hydrotreating for biofuel production. Source: Adapted from Davis et al. [31].

The benefits of this process include algae recycling and high-quality and compatible fuel production. However, the drawbacks are high energy use, high production costs, and environmental impacts for large-scale cultivation. The technology is in the early development stages and has TRL 4. Several institutions and companies (e.g., Chevron Lummus Global and Honeywell UOP) are involved in improving microalgae hydrotreating for biodiesel production. More information about the microalgae hydrotreating conversion process can be found in the earlier study by the National Renewable Energy Laboratory [31].

6.5 Case Study: Gasification with CCUS

The interest in the gasification process emerged in the 1970s during the energy crisis and later due to the need to reduce the environmental impacts of fossil fuel-based energy. Gasification converts feedstocks to syngas that can be used to generate power, hydrogen, hydrocarbon fuels, and chemicals. Currently, power generation through integrated gasification combined cycle (IGCC) is commercialized due to its higher efficiency and lower emissions compared to coal-based power generation. Globally, most in-operation gasification plants are in China, using different gasification reactors, such as down-draft, up-draft, entrained flow, and fluidized bed, with over 90% carbon conversion rates [10]. Table 6.3 provides detailed information about existing commercial gasifiers.

This case study focuses on converting syngas to liquid fuels and chemicals. Syngas from the gasification process has been commercially used to produce several fuels and chemicals, such as methanol, methane, substitute natural gas, and diesel. Raw syngas requires deep cleaning to adjust the H_2-to-CO ratio and reduce syngas impurities by removing particulate matter (e.g., ash and soot), sulfur, mercury, and heavy metals. For example, sulfur content should be reduced to less than 30 ppb (parts per billion) to avoid catalyst poisoning for proper CO–H_2 synthesis reactions. The key factors in the gasification process are feedstock selection based on the desired end product, gasification type, syngas cleanup and purification system, and plant size. The main aspects for feedstock selection are energy content

Table 6.3 Example of commercial gasification technologies.

Name	Description
British Gas Lurgi	Updraft, moving bed, slagging version of Lurgi
GE-Radiant Cooling	Slurry, entrained-flow, top fed, slagging
GE- full quench	Slurry, entrained flow, top fed, slagging
Lurgi Mark IV	Updraft, dry bottom moving bed, fuel fed from the top
Lurgi Multi-Purpose Gasifier	Entrained flow, top fed, multi-fuel, slagging
Shell SCGP	Dry, entrained flow, middle fed, water wall, slagging
Siemens	Dry, top fed, entrained flow, slagging

Source: Adapted from NETL [10].

(e.g., heating values), proximate analysis (e.g., volatile matters, fixed carbon, moisture, and ash), ultimate analysis (e.g., carbon, hydrogen, oxygen, and nitrogen), sulfur analysis, and ash composition.

Methanol and methane are two high-value products from syngas with high yields (over 95%). Methanol (CH_3OH) production from syngas is an established commercial process. The chemical reactions ($CO + 2H_2 \rightarrow CH_3OH$) are exothermic under high pressure, using a fixed-bed reactor and Cu–Zn/Al_2O_3 catalyst. This process includes methanol synthesis and WGS reaction. Methanol can be used for fuel or fuel additive and chemical production, such as methyl tert-butyl ether, formaldehyde, dimethyl ether, dimethyl amine, acetic acid, and olefins [10]. Fixed-bed and slurry reactors have been used for methanol production on a commercial scale (100 tons/day). Some leading producers of commercial-scale methanol plants are Lurgi, Synetix, Methanex, KBR, and Mitsubishi Gas Chemical. Also, the leading catalyst suppliers are Haldor Topsøe, BASF, Johnson Matthey, and Clariant.

Methane (substitute natural gas) production from syngas is a well-established commercial process. The chemical reactions ($CO + 3H_2 \rightarrow CH_4 + H_2O$) are exothermic, using selective catalytic methanation and nickel-based catalysts at around 350 °C. High pressure can increase the process yields. Several reactors have been tested, but fluidized-bed reactors have been commercialized by several companies, such as Lurgi, Clariant and Foster Wheeler, and Davy Technologies. Also, Ni-based catalysts are widely used in these processes.

FT process is another commercialized process for converting syngas to transportation liquid fuels (e.g., diesel), using different reactors (e.g., fixed-bed, fluidized-bed, or slurry reactors) and proper catalysts (e.g., iron or cobalt). The main benefits of these reactors include low capital cost, simplicity and ease of operation, and low catalyst consumption. Iron- and cobalt-based catalysts typically operate between 230 and 260 °C in a low-temperature FT process. Recently, co-based catalysts have been mainly used due to their higher stability and activity, low CO_2 selectivity, better integration with syngas, and higher attrition resistance in slurry reactors. The critical factors in these processes are reactor type, total and reactant pressure, temperature, and catalyst type. Several studies evaluated different catalysts for producing hydrocarbon fuels in a single-step process. Their results show that Co/Zeolite has 75% selectivity to diesel or jet fuel range compounds and solid wax-free products, and Fe+ZSM-5 has 60–70% selectivity to gasoline-range hydrocarbons and no wax production [32–34]. Also, Cu–Co and MoS_2 are widely examined for direct conversion of syngas to ethanol and higher alcohols, but their results show insufficient selectivity and yield for commercialization [35]. Since methanation, methanol synthesis, and FT processes are highly exothermic, efficient heat management is essential to maintain high process yield and prevent catalyst deactivation [10].

With an increasing demand for the gasification process for power, fuels, and chemicals production, CCUS strategies have become more crucial to capture CO_2 after the gasification process and before syngas conversion to other products. Currently, the main CCUS strategies are absorption, adsorption, membranes, and cryogenic processes [10]. The main challenges are: reducing energy requirements for carbon capturing and minimizing preparation requirements (e.g., pressure, target purity, temperature, and flow rate) to prevent carbon release to the environment by finding a suitable end use (e.g., enhanced oil recovery). As explained earlier, the primary methods for carbon capturing are pre-combustion (carbon is removed before combustion), post-combustion (carbon is removed after combustion), and oxy-fuel combustion (fuel is combusted with pure oxygen). Between these methods, pre-combustion can be a cost-competitive CCUS strategy for power and chemical production from high-pressure, oxygen-blown gasification processes (Figure 6.17) [10].

For fuels and chemical production (e.g., methanol), the amount of captured carbon is critical to make the raw syngas suitable for converting to end-use products. Therefore, different processes, such as WGS, have been used to convert CO and water to CO_2 and H_2 ($CO + H_2O \leftrightarrow CO_2 + H_2$) to produce syngas with a proper CO and H_2 ratio. These processes can improve conversion efficiency, productivity, and commercialization aspects. In addition, Selexol and Rectisol are the main commercial acid gas removal processes from high-pressure syngas, using physical solvents to capture carbon and hydrogen sulfide. In summary, the main factors for fuels and chemical production are the CO-to-H_2 ratio in the syngas and the effective carbon-capturing approach to improve the conversion process. Overall,

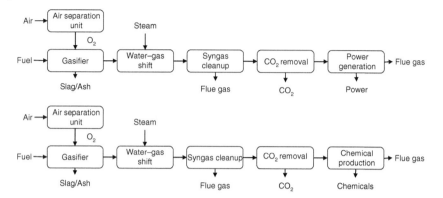

Figure 6.17 Process flow of gasification mixed with CCUS for power generation (top) and chemical production (bottom). Source: NETL [10].

gasification mixed with CCUS strategies for biofuel production is mainly dependent on the targeted products, and the major cost drivers are feedstock collection, transportation, and conversion processes.

Self-Check Questions

1. What are the main types of biofuels?
2. What are the most common biofuel production processes?
3. What is biogas?
4. What is anaerobic digestion?
5. What are the benefits and drawbacks of biogas production?
6. What are the main steps in the anaerobic digestion process? And what are the main factors?
7. What are the main parameters and challenges in algae anaerobic digestion?
8. What is biomethane?
9. What are the most used biomethane production technologies?
10. What are the main steps in the gasification process?
11. What are the main types of gasifiers?
12. What are the key parameters for evaluating the gasification process?
13. What are the applications of syngas from the gasification process?
14. What are the chemical reactions during the gasification process of organic materials (e.g., woody biomass or agriculture waste)?
15. What are the steps for biomethane production through gasification with catalytic methanation pathway?
16. What are the steps for biomethane production through gasification with biological methanation pathway?

17. What are the mature bioethanol production technologies?
18. What are the main steps during gasification syngas fermentation of lignocellulosic feedstocks for bioethanol production?
19. What are the chemical reactions during gasification syngas fermentation of lignocellulosic feedstocks?
20. What are the main steps during enzymatic fermentation of lignocellulosic feedstocks for bioethanol production?
21. What are the advantages and disadvantages of enzymatic fermentation of lignocellulosic feedstocks for bioethanol production?
22. How does enzymatic fermentation convert sugar and starch to bioethanol?
23. What are the benefits and drawbacks of enzymatic fermentation for bioethanol production?
24. What are the physical and chemical characteristic differences between biodiesel and diesel #2?
25. What are the benefits and drawbacks of biodiesel production?
26. What are the mature technologies for biodiesel production?
27. How does esterification convert fatty acids to biodiesel?
28. What are the key parameters for biodiesel production through esterification of fatty acids?
29. What are the steps during the hydrogenation of vegetable oil for biodiesel production?
30. How does the microbial lipid fermentation process produce biodiesel?
31. What are the benefits and drawbacks of the microbial lipid fermentation process for biodiesel production?
32. What are the steps and key parameters during the alcohol-to-jet fuel process?
33. What are the benefits and drawbacks of the alcohol-to-jet fuel process?
34. What are the steps for biofuel production through pyrolysis and upgrading processes?
35. What are the steps for biofuel production through hydrothermal liquefaction and upgrading processes?
36. What are the chemical reactions for biofuel production through gasification-FT with and without CCUS?
37. What are the steps for biofuel production through gasification-FT mixed with hydrogen enhancement?
38. What is the transesterification process?
39. What are the steps for biodiesel production from microalgae through the transesterification process?
40. What is the hydrotreating conversion process?
41. What are the main CCUS strategies?
42. What are the primary carbon-capturing methods?
43. What are the challenges of using CCUS strategies?

References

1 USDA (2022). Creating Energy and Fuel from Biomass.

2 DOE (2022). Biomass Resources.

3 Rogers, J.N., Stokes, B., Dunn, J. et al. (2017). An assessment of the potential products and economic and environmental impacts resulting from a billion ton bioeconomy. *Biofuels, Bioproducts and Biorefining* 11 (1): 110–128.

4 Grimalt-Alemany, A., Skiadas, I.V., and Gavala, H.N. (2018). Syngas biomethanation: state-of-the-art review and perspectives. *Biofuels, Bioproducts and Biorefining* 12 (1): 139–158.

5 U.S. EPA (2019). Methane Emissions in the United States: Sources, Solutions & Opportunities for Reductions.

6 Kumar, M., Dutta, S., You, S. et al. (2021). A critical review on biochar for enhancing biogas production from anaerobic digestion of food waste and sludge. *Journal of Cleaner Production* 305: 127143.

7 Ward, A.J., Lewis, D.M., and Green, F.B. (2014). Anaerobic digestion of algae biomass: a review. *Algal Research* 5: 204–214.

8 Quan, L.M., Kamyab, H., Yuzir, A. et al. (2022). Review of the application of gasification and combustion technology and waste-to-energy technologies in sewage sludge treatment. *Fuel* 316: 123199.

9 Sharma, I., Rackemann, D., Ramirez, J. et al. (2022). Exploring the potential for biomethane production by the hybrid anaerobic digestion and hydrothermal gasification process: a review. *Journal of Cleaner Production* 362: 132507.

10 NETL (2022). *Guidelines/Handbook For The Design of Modular Gasification Systems.*

11 Patuzzi, F., Basso, D., Vakalis, S. et al. (2021). State-of-the-art of small-scale biomass gasification systems: an extensive and unique monitoring review. *Energy* 223: 120039.

12 Chew, K.R., Leong, H.Y., Khoo, K.S. et al. (2021). Effects of anaerobic digestion of food waste on biogas production and environmental impacts: a review. *Environ Chem Lett* 19 (4): 2921–2939.

13 US DOE BETO (2015). *Enhanced Anaerobic Digestion and Hydrocarbon Precursor Production from Sewage Sludge.*

14 Ashok, J., Pati, S., Hongmanorom, P. et al. (2020). A review of recent catalyst advances in CO2 methanation processes. *Catalysis Today* 356: 471–489.

15 DOE (2022). *Alternative Fuels Data Center: Ethanol Fuel Basics.*

16 US DOE (2022). *Ethanol Benefits and Considerations.*

17 Melendez, J.R., Mátyás, B., Hena, S. et al. (2022). Perspectives in the production of bioethanol: a review of sustainable methods, technologies, and bioprocesses. *Renewable and Sustainable Energy Reviews* 160: 112260.

18 Kennes, D., Abubackar, H.N., Diaz, M. et al. (2016). Bioethanol production from biomass: carbohydrate vs syngas fermentation. *Journal of Chemical Technology & Biotechnology* 91 (2): 304–317.

19 Tan, E.C.D., Talmadge, M., Dutta, A. et al. (2015). *Process Design and Economics for the Conversion of Lignocellulosic Biomass to Hydrocarbons via Indirect Liquefaction. Thermochemical Research Pathway to High-Octane Gasoline Blendstock Through Methanol/Dimethyl Ether Intermediates (No. NREL/TP-5100-62402).* Golden, CO (United States): National Renewable Energy Lab.

20 Alleman, T.L., McCormick, R.L., Christensen, E.D. et al. (2016). *Biodiesel Handling and Use Guide*, 5e. Golden, CO (United States): National Renewable Energy Lab (NREL).

21 Pechout, M., Kotek, M., Jindra, P. et al. (2019). Comparison of hydrogenated vegetable oil and biodiesel effects on combustion, unregulated and regulated gaseous pollutants and DPF regeneration procedure in a Euro6 car. *Science of The Total Environment* 696: 133748.

22 Wang, W.-C., Tao, L., Markham, J., Zhang, Y., Tan, E., Batan, L., Warner, E., and Biddy, M. (2016). Review of biojet fuel conversion technologies.

23 Jones, S.B., Valkenburt, C., Walton, C.W., Elliott, D.C., Holladay, J.E., Stevens, D.J., Kinchin, C., and Czernik, S. (2009). Production of gasoline and diesel from biomass via fast pyrolysis, hydrotreating and hydrocracking: a design case.

24 Hansen, S., Mirkouei, A., and Diaz, L.A. (2020). A comprehensive state-of-technology review for upgrading bio-oil to renewable or blended hydrocarbon fuels. *Renewable and Sustainable Energy Reviews* 118.

25 Snowden-Swan, L.J., Li, S., Jiang, Y. et al. (2022). *Wet Waste Hydrothermal Liquefaction and Biocrude Upgrading to Hydrocarbon Fuels: 2022 State of Technology. No. PNNL-32731.* Richland, WA (United States): Pacific Northwest National Lab (PNNL).

26 Jones, S.B., Zhu, Y., Anderson, D.B. et al. (2014). *Process Design and Economics for the Conversion of Algal Biomass to Hydrocarbons: Whole Algae Hydrothermal Liquefaction and Upgrading (No. PNNL-23227).* Richland, WA (United States): Pacific Northwest National Lab.(PNNL).

27 Snowden-Swan, L.J., Billing, J.M., Thorson, M.R. et al. (2021). *Wet Waste Hydrothermal Liquefaction and Biocrude Upgrading to Hydrocarbon Fuels: 2020 State of Technology.* Pacific Northwest National Laboratory https://doi.org/10.2172/1863608.

28 Laurens, L.M.L., Quinn, M., Van Wychen, S. et al. (2012). Accurate and reliable quantification of total microalgal fuel potential as fatty acid methyl esters by in situ transesterification. *Analytical and Bioanalytical Chemistry* 403 (1): 167–178.

29 Chen, H., Wang, X., and Wang, Q. (2020). Microalgal biofuels in China: the past, progress and prospects. *Gcb Bioenergy* 12 (12): 1044–1065.

30 Salam, K.A., Velasquez-Orta, S.B., and Harvey, A.P. (2016). A sustainable integrated in situ transesterification of microalgae for biodiesel production and associated co-product-a review. *Renewable and Sustainable Energy Reviews* 65: 1179–1198.

31 Davis, R., Kinchin, C., Markham, J. et al. (2014). *Process Design and Economics for the Conversion of Algal Biomass to Biofuels: Algal Biomass Fractionation to Lipid- and Carbohydrate-Derived Fuel Products (No. NREL/TP-5100-62368).* Golden, CO (United States): National Renewable Energy Lab (NREL).

32 Kibby, C., Jothimurugesan, K., Das, T. et al. (2013). Chevron's gas conversion catalysis-hybrid catalysts for wax-free Fischer–Tropsch synthesis. *Catalysis Today* 215: 131–141.

33 Zhang, Q., Cheng, K., Kang, J. et al. (2014). Fischer–Tropsch catalysts for the production of hydrocarbon fuels with high selectivity. *ChemSusChem* 7 (5): 1251–1264.

34 Gangwal, S.K. and McCabe, K. (2015). *Small-Scale Coal-Biomass to Liquids Production Using Highly Selective Fischer-Tropsch Synthesis.* Durham, NC (United States): Southern Research Institute.

35 Subramani, V. and Gangwal, S.K. (2008). A review of recent literature to search for an efficient catalytic process for the conversion of syngas to ethanol. *Energy & Fuels* 22 (2): 814–839.

7

Hydrogen Production

This chapter provides the existing low-emission technologies for hydrogen production (Figure 7.1). Hydrogen is an energy carrier and carbon-free fuel, consisting of one proton and one electron that can store and deliver energy from other sources [1]. Hydrogen has the potential to be a valuable energy source due to its benefits compared to fossil fuels (e.g., high energy content and lower greenhouse gas [GHG] emissions). The energy content of hydrogen is significantly higher (around 120 MJ/kg) compared to other energy sources, such as gasoline (around 44 MJ/kg). Hydrogen has several applications and can be used across several sectors, such as clean energy systems for power generation, using hydrogen-powered turbines or fuel cells, and fuel and chemical production through hydrogenation.

The science behind hydrogen production mainly involves separating it from other elements, and the most common methods are steam-methane reforming, electrolysis, and water splitting with power (Figure 7.2). Currently, hydrogen production from fossil resources is the most cost-effective approach for large-scale production. Additionally, their byproducts (e.g., solid carbon) have applications in various sectors. The existing commercial methods for hydrogen production from fossil fuel resources are coal gasification and natural gas conversion that can be combined with carbon capture, utilization, and sequestration (CCUS) strategies to reduce GHG emissions. Hydrogen production methods from biomass feedstocks and wastes include biogas reforming and fermentation of waste streams coupled with CCUS. The byproducts of these processes can be used in water treatment, power generation, and chemical production. Hydrogen production methods through water splitting can use low- or high-temperature electrolysis, and the required power for splitting can be provided from nuclear or renewable energy sources, such as solar, wind, and geothermal.

Hydrogen is used mainly for ammonia production in the fertilizer industry, methanol production in the chemical industry, and oil refinery in the transportation industry. The main challenges are the production cost, handling, and safety aspects. Recent studies reported that the United States produces around

Net-Zero and Low Carbon Solutions for the Energy Sector: A Guide to Decarbonization Technologies,
First Edition. Amin Mirkouei.
© 2024 John Wiley & Sons, Inc. Published 2024 by John Wiley & Sons, Inc.

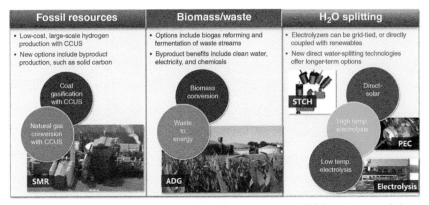

ADG: Anaerobic digester gas; CCUS: Carbon capture, utilization, and storage; PEC: Photoelectrochemical; SMR: Steam; methane reforming; STCH: solar thermochemical hydrogen

Figure 7.1 Classification of low-emission hydrogen production technologies. Source: U.S. Department of Energy (public domain) [1].

10 million tons of hydrogen per year, mainly by steam reforming natural gas, and the primary demands are fuel and chemical production [1]. In addition, hydrogen can be an effective solution to enable near-zero emission processes and address national challenges in mitigating global warming potentials due to carbon emissions from fossil fuels. The high production cost, storing, and transporting are the main challenges of hydrogen production, especially via net-zero carbon pathways. According to the U.S. DOE, hydrogen production costs around $5/kg from renewable sources. Reducing production costs can address national priorities using local resources, such as clean energy security, environmental impact mitigation, and job creation. Particularly, it can help to reduce over 15% of carbon emissions by 2050 and create over 500,000 jobs by 2030 [1].

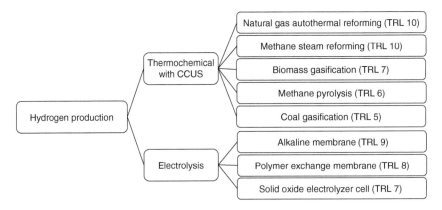

Figure 7.2 Classification of hydrogen production technologies

An overview of key low-emission solutions for hydrogen production is presented in this chapter. Particularly, this chapter provides eight technologies with high technology readiness levels, along with a case study about hybrid thermochemical hydrogen production.

7.1 Thermochemical with CCUS

Fossil fuels, especially natural gas and coal, can be reformed to generate hydrogen from their hydrocarbon molecules. Currently, most hydrogen (approximately 95%) have produced from natural gas and other fossil fuels in the United States through steam-methane reforming of natural gas (91%) and partial oxidation of natural gas via coal gasification (4%) that can be mixed with CCUS methods to capture carbon and reduce the emissions [2]. Thermochemical processes use high-temperature methods to drive chemical reactions and produce hydrogen. The basic science involves breaking down the chemical compounds (e.g., CH_4 and H_2O) into CO and H_2, as well as separation and purification. There are several thermochemical methods for hydrogen production, such as methane steam reforming, natural gas autothermal reforming, gasification, pyrolysis, and thermochemical water splitting. Globally, methane steam reforming of natural gas, coal gasification, and electrolysis can produce approximately 76%, 22%, and 2% hydrogen [2, 3].

The key parameters for hydrogen production via thermochemical processes include the type of feedstock (e.g., water or methane), conversion reactor (e.g., fluidized- or fixed-bed), catalyst, required energy and heat source (e.g., nuclear, renewable, or fossil fuels), as well as separation and purification methods. The produced hydrogen from thermochemical processes needs further purification to achieve the required specifications (e.g., purity) for the commercial scale (Table 7.1). Currently, pressure swing adsorption is the most common commercial

Table 7.1 Hydrogen quality requirements.

Specification	Amount
Hydrogen	99.97%
Other gases	300 ppm
Helium	300 ppm
Nitrogen and argon	300 ppm
Methane	100 ppm
Water	5 ppm
Oxygen	5 ppm
Particle size	10 μm

method to achieve the required hydrogen quality, using a molecular sieve by removing other gases (e.g., nitrogen, oxygen, CO, and H_2S). The pressure swing adsorption process includes five steps: adsorption, co-current depressurization, purge, and counter-current depressurization.

The main benefits of thermochemical processes mixed with CCUS include high efficiency, low emissions, and feedstock flexibility. The main drawbacks are high requirements (e.g., energy, temperature, and pressure) and high capital and operational costs. The environmental impacts are highly dependent on feedstock type and energy source. Some of the leading producers of hydrogen through thermochemical processes are Air Liquide, Siemens, and Linde.

7.1.1 Methane Steam Reforming (TRL 10)

Globally, methane steam reforming, using natural gas, has the highest share of hydrogen production among the existing methods due to its high capacity and efficiency. Steam reforming is an endothermic process that requires heat, and can be used to generate hydrogen from other sources, such as gasoline, ethanol, and propane. Methane steam reforming is a mature process that can generate hydrogen from methane, using high-temperature steam (700–1,000 °C). In this process, methane reacts with steam under high pressure (up to 360 psi) in the presence of catalysts (Figure 7.3). In the water–gas shift process, CO and steam are reacted in the presence of a catalyst to produce CO_2 and hydrogen. The chemical

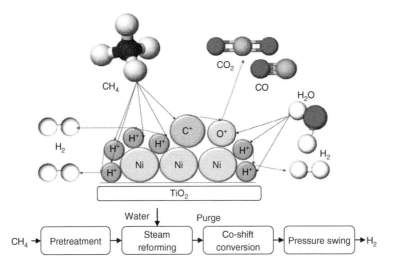

Figure 7.3 Schematic of methane steam reforming process for hydrogen production in the presence of Ni with the support of TiO_2. Source: Boretti and Banik [4]/John Wiley & Sons.

reaction of the methane steam reforming process for hydrogen production is as follows:

$$CH_4 + 2H_2O + heat \rightarrow CO_2 + 4H_2$$

The main components of methane steam reforming are: (i) the reformer for steam and methane reaction, (ii) heat source for the endothermic process, (iii) the catalyst and its support (e.g., nickel on alumina), and (iv) the separation processes, such as pressure swing adsorption. The key parameters of this process include reaction temperature and pressure, catalyst performance, and steam and carbon ratio. Catalyst performance is critical to process yields and efficiency. Some of the main reasons for catalyst deactivation are carbon deposition, thermal degradation, and sulfur poisoning.

The benefits of methane steam reforming are reliability and high efficiency for large-scale hydrogen production from natural gas, biogas, and coal. The byproducts of this process (e.g., CO) can be used in different applications to increase efficiency and reduce costs. The main drawbacks include: (i) GHG emissions (e.g., CO_2, SO_2, and NO_x) and environmental impacts of using fossil fuel-based resources (e.g., natural gas or coal), (ii) energy-intensive processes, and (iii) expensive separation processes. The major cost drivers are feedstock cost, energy use, and separation costs. Air Liquide, Linde, and Air Products are the leaders in producing hydrogen through methane steam reforming process. Latest studies show that hydrogen production through biomass steam reforming with 74–85% energy efficiency that can produce 40–130g/kg feedstock, which costs $1.83–$2.35 per kg of hydrogen [5].

Methane steam reforming for hydrogen production from fossil fuels releases GHG emissions that can be minimized with CCUS technologies. One of the key CCUS methods is to capture CO_2 from high-pressure syngas after the first conversion process before hydrogen separation, similar to pre-combustion carbon removal processes in coal power plants. The main benefits of this capturing pathway are high CO_2 concentration and low capital costs, but the main drawback is the low CO_2 capturing rate (up to 60%) [6]. Another main CCUS method is to capture CO_2 from low-pressure streams after hydrogen separation, similar to post-combustion in power plants. The key benefit is high capturing rare (around 90%); however, it requires high capital costs. The latest techno-economic studies show that the levelized cost of hydrogen production increases between 18% and 45% when the steam reforming process is integrated with CCUS technologies. Currently, two hydrogen production plants in the United States and Canada use methane steam reforming combined with CCUS, which are Air Product in Texas, the United States, and Shell in Alberta, Canada. Both plants use pre-combustion carbon-capturing technology, and their latest results show that they can capture over a million metric tons of CO_2 per year. More detailed information has been provided in [6].

7.1.2 Natural Gas Autothermal Reforming (TRL 10)

Natural gas autothermal reforming is a mature process for hydrogen production, using a combination of reforming processes (e.g., steam reforming, partial oxidation, and autothermal reforming), along with other processes such as purification and cleanup for removing impurities (Figure 7.4). The United States produces over 94% of hydrogen through natural gas reforming, particularly thermal processes such as partial oxidation or steam-methane reforming, since methane is the primary gas source in natural gas. The main drawback of hydrogen production from natural gas (or fossil fuels) is the high GHG emission (900 Mt CO_2 [8]).

The autothermal reforming process converts natural gas to syngas (e.g., H_2, CO, and CO_2) through oxidizing natural gas using oxygen in the presence of a catalyst (e.g., nickel-based materials). Then CO_2 is removed to produce H_2-rich gas. The science behind this process involves chemical reactions, kinetics, and thermodynamics. Partial oxidation is an exothermic process that releases heat and can generate hydrogen from methane in natural gas with limited oxygen (or air). The partial oxidation of the methane reaction is as follows:

$$CH_4 + \frac{1}{2}O_2 \rightarrow CO + 2H_2 + heat$$

The main parameters to improve the output and efficiency of the natural gas autothermal reforming process are: (i) reactor design and configuration (e.g., temperature and pressure), (ii) natural gas, oxygen flow, and mixing rates; and (iii) catalyst type, bed design, and composition. Partial oxidation is faster than steam reforming, but it generates less hydrogen. The main benefits of the autothermal reforming process are: (i) cost-effective method compared to other methods (e.g., gasification or electrolysis), (ii) high energy efficiency (up to 80%), and (iii) high purity H_2 level that can address several industrial needs. The main drawbacks are: (i) the high capital cost of the oxygen production process, which can be addressed in large-scale hydrogen production; (ii) an energy-intensive process; and (iii) high

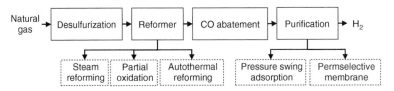

Figure 7.4 Block diagram of hydrogen production through natural gas autothermal reforming. Source: Adapted from Ciambelli et al. [7].

GHG emissions and environmental impacts that can be addressed by mixing this process with CCUS strategies. The major cost drivers are natural gas price and energy use. The United States, China, and Russia are the largest producer of hydrogen through natural gas autothermal reforming, and some of the leading companies are Air Products and Chemicals, Linde, Haldor Topsoe, and Thyssenkrupp Industrial Solutions.

The main differences between natural gas autothermal reforming and methane steam reforming include: (i) autothermal reforming process uses both oxygen and steam, but methane reforming uses only steam; (ii) autothermal process requires higher temperatures (up to 1,000 °C) and pressures (up to 30 bar); (iii) autothermal process produces H_2, CO, CO_2, and water, but methane reforming produces H_2, CO_2, and a small amount of CO; (iv) autothermal process is more energy efficient, has a higher yield, and requires less heat; (v) autothermal process requires catalysts for both oxidation and steam reforming, but methane reforming needs catalysts only for steam reforming reaction; and (vi) it is easier to mix CCUS strategies with autothermal process due to the gas mixture, such as higher CO_2 concentrations. In summary, the autothermal reforming process is more complex and requires more energy and catalyst; however, it has higher hydrogen yields and effective CCUS integration.

7.1.3 Biomass Gasification (TRL 7)

Biomass feedstocks can be used to generate hydrogen through various thermochemical processes, such as gasification and pyrolysis, that have less environmental impact and release low GHG emissions compared to fossil fuel sources (Figure 7.5). Gasification is a mature, high-temperature process (above 700 °C) that can produce syngas (a mixture of H_2, CO, CO_2, and other gases) from biomass feedstocks and organic wastes with H_2 and CO. The two main types of biomass feedstocks for hydrogen production are energy crops (e.g., poplar and

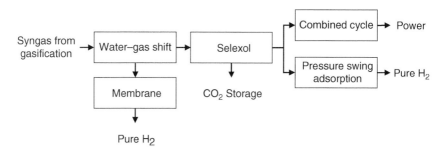

Figure 7.5 Block diagram of hydrogen production from gasification syngas using membrane or pressure swing adsorption. Source: Adapted from NETL [2].

switchgrass) and biogas from organic residues (e.g., animal wastes and municipal solid wastes) after the anaerobic digestion process. The chemical reaction of the biomass gasification process for hydrogen production is as follows:

$$C_6H_{9.5}O_{4.4} + 7.6H_2O + heat \rightarrow 6CO_2 + 12.4H_2$$

The key parameters in gasification are biomass type, operation conditions (e.g., reactor type, temperature, residence time, catalyst), gasification agent (air or steam), syngas yield, and hydrogen content. The tar production during the gasification process is one of the major issues that can limit the commercialization of hydrogen generation through this method. Catalytic gasification can address some issues (e.g., tar production) and improve the process yield.

The main benefits of hydrogen production through biomass gasification include: (i) biomass and waste management, (ii) lower environmental impacts compared to fossil fuels due to the use of renewable and local resources, (iii) a flexible and diverse source of hydrogen from a variety of feedstocks, and (iv) the need for high CO_2 storage. The major drawbacks are feedstock availability throughout the year, pretreatment processes (e.g., size reduction), syngas cleaning to reduce impurities (e.g., tar, acids, and particulates), and the high cost of storing and transporting large quantities. The major cost drivers are feedstock (over 80%), capital cost (around 12%), and operation cost (around 5%), including energy use, gas cleaning and condition, and hydrogen purification and compression. Latest studies show that hydrogen production through biomass gasification with 30–60% energy efficiency can produce 40–190 g/kg feedstock, which costs $1.77–$2.05 per kg of hydrogen [5].

There are several strategies to improve the process efficiency and reduce environmental impacts, such as mixing with CCUS strategies and using a heat recovery system that can reuse the heat from the gasification process for various applications, such as dewatering the feedstocks. In addition, various approaches can be used to reduce the levelized cost, including (i) combining processes into a single-step process, such as water–gas shift and pressure swing adsorption separation; (ii) using flexible feedstocks or other gas cleanup methods; (iii) improving heat integration and catalyst durability; and (iv) improving hydrogen yields through catalyst selectivity. Currently, China is the largest producer of hydrogen from biomass gasification, and some of the leading technology developers are Hitachi Zosen Corporation, Foster Wheeler, and Mitsubishi Heavy Industries. More detailed information on the effects of gasification parameters on hydrogen production can be found in [9].

7.1.4 Methane Pyrolysis (TRL 6)

As explained in Chapter 6, the pyrolysis process is a thermochemical process similar to gasification, but it operates under lower temperatures (around 500 °C) in the absence of oxygen. Recent studies investigated hydrogen production from various biomass feedstocks by producing hydrogen-rich gases (e.g., CH_4) via pyrolysis. The gas yield from pyrolysis was around 20–30%, containing a volumetric composition of up to 55% hydrogen, depending on biomass types [10]. In addition, methane pyrolysis can convert methane to hydrogen and carbon ($CH_4 + heat \rightarrow C + 2H_2$) through a high-temperature (endothermic) and non-catalytic process (Figure 7.6).

The main components and parameters that can affect the process yields and efficiency are: (i) feedstock and reactor type, (ii) process specifications (temperature and residence time), (iii) carbon and hydrogen recovery systems, and (iv) methane flow rate. The benefits of this process are: (i) low GHG emissions, (ii) non-catalytic process, (iii) no requirements for purification steps due to high purity hydrogen, and (iv) potential to integrate with CCUS strategies for carbon production. The drawbacks include tar formation, high energy use, and high capital and operational costs. Latest studies show that hydrogen production through biomass pyrolysis with 35–50% energy efficiency can produce 25–65 g/kg feedstock that costs $1.59–$2.20 per kg of hydrogen [5]. Currently, Germany is the leading producer of hydrogen from methane pyrolysis, and some of the leading technology developers are Thyssenkrupp AG, Fraunhofer IKTS, and Hydrogenious LOHC.

Figure 7.6 Schematic of hydrogen production through methane pyrolysis. Source: Lott et al. [11]/John Wiley & Sons/CC BY 4.0.

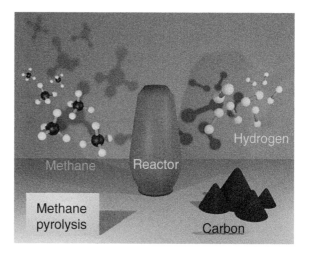

7.1.5 Coal Gasification (TRL 5)

Gasification process is a high-temperature (700–1,200 °C) process that can convert carbon-based sources (e.g., coal) to synthetic gas (syngas) using various agents (e.g., air, steam, or oxygen). Syngas needs to be purified from particles, substances, and contaminants through gas cleanup processes. Coal-based hydrogen production via the gasification process gained interest recently due to high hydrogen yield, reduced GHG emissions, and feedstock flexibility; however, this process requires high energy (e.g., power) to provide pure oxygen as the gasification agent. The chemical reaction of the coal gasification process for hydrogen production is as follows:

$$C_{8.2}H_4O + 15.4H_2O + heat \rightarrow 8.2CO_2 + 17.4H_2$$

The primary science principles of gasification include thermodynamics and kinetics, particularly Gibbs free energy charge for efficiency analysis and water–gas shift reactions. The key parameters that affect the process efficiency are coal type, process temperature and pressure, and gas mixture composition. The main components of this process are the gasification reactor, water–gas shift reactor, and gas purification unit. The major drawback of hydrogen production from coal gasification is the negative environmental impacts (e.g., high carbon emissions) that can be addressed by integrating with CCUS strategies (Figure 7.7). Other drawbacks are the high cost of storing and transporting large quantities. The latest study by the National Energy Technology Laboratory compared performance efficiency, costs, and emissions of hydrogen production through

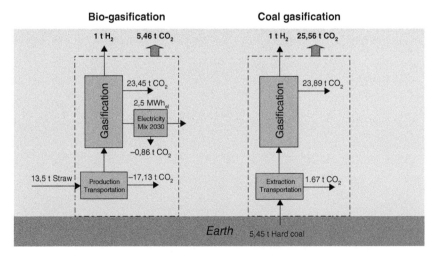

Figure 7.7 Environmental impact assessment of a ton hydrogen production through the gasification of biomass or coal. Source: Machhammer et al. [12]/John Wiley & Sons.

coal gasification with and without carbon capture. The results of coal gasification mixed with carbon capturing show the levelized cost of hydrogen is higher (around $0.34 per kg of hydrogen) with 92% carbon capture [13].

The National Energy Technology Laboratory also investigated hydrogen production through coal gasification without power export. They concluded that this conversion pathway has high potential due to its simplicity by eliminating the need for hydrogen-fired gas turbines and eliminating the need to compress air separation units [14]. Their results show that the overall efficiencies (HHV%) with and without carbon capturing, using a GE gasifier, were approximately 63.7% and 59%, respectively. The major cost drivers in coal gasification for hydrogen production are feedstock (i.e., coal), capital costs, and operation costs. Currently, China is the largest producer of hydrogen through coal gasification for local use. Further information about coal gasification for hydrogen production can be found in [15].

7.2 Electrolysis

Electrolysis of water is one of the commercialized electrochemical methods for hydrogen production. Water electrolysis uses two electrodes in a solution of water and electrolyte (e.g., NaOH or KOJ) that are connected to the power supply to decompose (split) water molecules into hydrogen and oxygen at the cathode and anode, respectively. The science of electrolysis is based on Faraday's law (the relationship between the amount of power used to drive chemical reactions) under 50 °C and 1–100 bar pressure with an efficiency of around 70%. The overall electrolysis reaction is as follows:

$$2H_2O + \text{electricity} + \text{heat} \rightarrow 2H_2 + O_2$$

The main components of electrolysis are electrodes, power supply, electrolysis cell, and electrolyte solution. The main parameters that can affect the process efficiency include electrodes' surface area, applied voltage and current, electrolyte type and concentration, and process configurations (e.g., pressure and temperature). Earlier studies show that higher electrolyte concentrations and higher voltages and currents can enhance reaction rate and hydrogen production. However, it can also increase side reactions, electrode corrosion, and process temperature.

The key benefits of water electrolysis for hydrogen or power production are simplicity, low temperature, zero carbon emissions, high hydrogen purity, and oxygen byproducts. The drawbacks are process requirements (e.g., high pressure), energy storage challenges, low efficiencies, and high capital costs. Recent results show that water electrolysis for hydrogen production has 55–80% energy efficiency, can produce 111 g/kg feedstock, and costs $4.15–$10.30 per kg

Table 7.2 Mature hydrogen production method comparison.

Method	Resource	Efficiency (%)	Cost ($/kg)	Mechanism
Autothermal reforming	Natural gas or methane	60–75	1.51 (94% carbon capture)	Exothermic reaction, high temperature (950 °C with catalyst and over 1,200 °C without catalyst), and pressure (5.5–6 MPa)
Steam reforming	Methane, oil, or coal	74–85	1.06–1.54 (96% carbon capture)	Endothermic reaction, high temperature, and pressure (3.5 MPa)
Gasification	Biomass or coal	35–50	2.58–2.92 (92% carbon capture)	Endothermic/ exothermic reaction, high temperature (500–1,300 °C)
Pyrolysis	Biomass	35–50	3.44 (92% carbon capture)	High temperature (300–650 °C)
Electrolysis	Water	60–80	4.15–10.30	Alkaline electrolysis systems

Source: Adapted from Alptekin and Celiktas [9], NETL [13], Midilli et al. [15].

of hydrogen [5]. Table 7.2 compares different hydrogen production methods and their attributes.

Globally, Europe (60%) and China (35%) are the largest producers of electrolysis technologies, and some of the leading companies are Nel Hydrogen, Thyssenkrupp, and ITM Power. Electrolysis is in early development and needs further efficiency and durability improvements. Other hydrogen production technologies that use water are thermolysis, photoelectrolysis, and biophotolysis, with up to 50%, 14%, and 15% energy efficiency, respectively [5]. These technologies have several benefits, but they are in their early research and development stage and require further improvement to address their main drawbacks, such as low efficiency and high cost.

7.2.1 Alkaline Membrane (TRL 9)

Alkaline electrolysis process is a long-established and well-mature technology that uses alkaline electrolyte solutions (e.g., KOH and NaOH) and operates

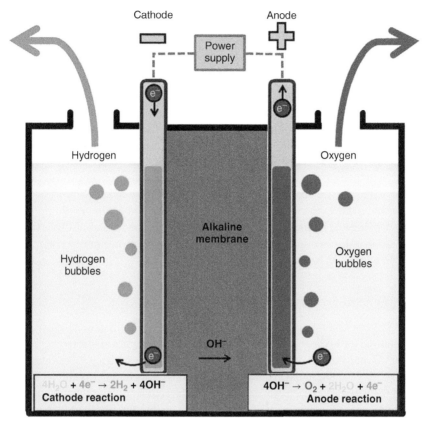

Figure 7.8 Schematic of electrolysis process with alkaline membrane for hydrogen production. Source: Pozio et al. [16]/John Wiley & Sons.

at 100–150 °C with stack voltage efficiency between 60 and 80% (Figure 7.8). According to the latest studies, the alkaline electrolysis process dominated over 60% of the installed capacity in 2020 [17]. This technology has a dominant place in the current market and is a low-cost, mature process at scale due to its good durability. The main parameters affecting the process efficiency are electrolyte concentration, current density (current amount per unit area), temperature, and pressure. For example, a higher current density can increase the hydrogen production rate, but it can increase energy use and reduce the electrolysis cell lifespan. Also, the process pressure depends on the used electrolyte and cell design. Generally, the pressure is around atmospheric pressure, but it can be elevated in order to improve the process efficiency and hydrogen production rate. Subsequently, the high-pressure process requires stronger and more costly materials.

The main benefits of alkaline electrolysis for hydrogen production include low capital costs due to low material costs (compared to other electrolysis technologies), long lifetime, high hydrogen production, compact design, and high energy efficiency (over 80%) [16]. The drawbacks include low hydrogen purity, low partial load range, low current density, and complex KOH (potassium hydroxide) handling. The commonly used electrolyte solution is KOH, and the catalysts are common metals. In addition, KOH is corrosive to metals in moist air. The latest techno-economic analysis in China reported that alkaline electrolyzers cost between $750 and $1,300 per kW. The major cost drivers are anode (25%), cathode (24%), and structural rings (14%).

Recent studies show that large cells, power from renewable sources (e.g., solar and wind), simplified cell design, and higher power density can reduce the total cost of hydrogen production through alkaline electrolysis. Currently, China, the United States, and Germany are the largest producers of alkaline electrolysis technology, and some of the leading producers are Zhongdian Hydrogen Equipment, Proton OnSite, and Siemens Energy. One of the largest alkaline electrolysis plants in operation is the Cachimayo plant in Peru. More detailed information about alkaline electrolysis can be found in [18].

7.2.2 Polymer Exchange Membrane (TRL 8)

The polymer exchange membrane electrolysis process is the second most common method for hydrogen production from water. It operates at low temperatures (70–90 °C) and various pressures, and its stack voltage efficiency is between 65% and 80%, very similar to the alkaline electrolyzer. This technology dominated over 30% of the installed capacity in 2020 [17]. The basic science is to separate oxygen and hydrogen in water, using a proton-exchange membrane. Similar to other electrolysis processes, the key components of this process are anode, cathode, membrane, and electrolyte. During this process, the anode produces oxygen gas and proton from water, then protons pass the membrane to the cathode, where protons and electrons from external power sources produce hydrogen gas.

Proton-exchange membrane electrolysis uses electrode catalysts (e.g., platinum), bipolar plates (e.g., titanium), and membrane materials, which make it more expensive compared to the alkaline electrolysis process (Figure 7.9). Nafion-based membranes are the most popular proton-exchange membranes due to their high ionic conductivity, thermostability, and durability at low temperatures. The commonly used catalysts are rare metals. Similar to alkaline electrolysis, the main parameters are process temperature and pressure, current density, and electrolyte composition. Higher temperature and pressure can improve the performance of electrolysis cells and increase the hydrogen production rate.

The benefits of the proton-exchange membrane electrolysis process are high proton conductivity, high efficiency, high hydrogen purity, modular design, low process footprint, pure water solution (no alkaline), and fast response. The

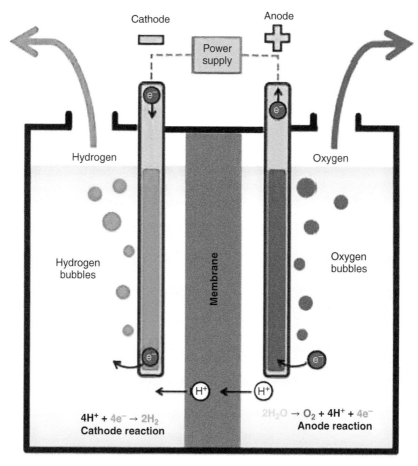

Figure 7.9 Schematic of polymer electrolyte membrane electrolyzer. Source: U.S. Department of Energy (public domain) [6].

drawbacks are high material costs, supply chain challenges for precious metals, shorter lifespan than alkaline electrolysis, limited capacity, and low durability. The major cost drivers are bipolar plates (51%) and membrane electrode assembly manufacturing (10%). Currently, hydrogen production, using proton-exchange membrane electrolysis costs around $5.12 per gallon gas equivalent of hydrogen [6]. The largest producers of this technology are in Europe, such as Nel Hydrogen and Siemens Energy. Air Liquide developed one of the largest proton-exchange membrane electrolysis plants in operation in Bécancour, Canada. More detailed information about membrane-based electrolysis for hydrogen production can be found in [17, 19].

7.2.3 Solid Oxide Electrolyzer Cell (TRL 7)

Solid oxide electrolyzer is one of the mature technologies for hydrogen production from water (Figure 7.10). This technology operates by passing an electric current through a solid oxide electrolyte at very high temperatures (700–800 °C) for splitting water, and its efficiency is around 100%. The solid oxide electrolyte is usually a ceramic material with high ionic conductivity. The main components are the anode, cathode, solid oxide electrolytes, current collectors to connect electrodes to an external circuit, and seals to avoid gas leaks.

The key parameters affecting solid oxide electrolyzer performance include process temperature and pressure, current density, and electrolyte composition. Particularly, high temperature and pressure can increase the reaction rates and efficiency. In this process, current density and electrolyte composition can affect the electrochemical reactions, ion conductivity, and stability. The main benefits are high efficiency (up to 80%), flexibility to produce hydrogen and power as fuel cells, and a long lifespan. However, the drawbacks are high cost, high-temperature requirements, and carbon deposition that can decrease process efficiency. The solid oxide electrolyzer has lower maturity compared to alkaline and polymer

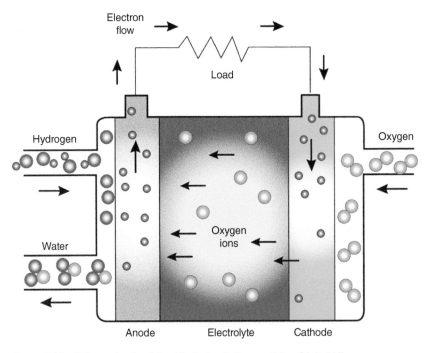

Figure 7.10 Schematic of solid oxide fuel cell. Source: Wiley Digital Library.

exchange membrane electrolysis processes due to the operation requirements (e.g., high temperature). The extra heat can be reused in other processes.

The major cost drivers are energy consumption and high-performance materials that can handle high temperatures, pressures, and corrosive process. Currently, hydrogen production using solid oxide electrolyzer costs around $4.95 per gallon gas equivalent of hydrogen [6]. This technology can produce hydrogen without carbon emissions if the power is provided from renewable sources (e.g., wind and solar). The main negative environmental impact is the significant amount of water. Currently, several companies produce solid oxide electrolyzers, such as Siemens Energy, Mitsubishi Power, ITM Power, and McPhy Energy.

7.3 Case Study: Hybrid Thermochemical Hydrogen Production

Hydrogen is an exceptional energy carrier that can be used in various sectors (e.g., transportation and manufacturing) to mitigate GHG emissions, similar to renewable power sources, such as solar and wind. Fuel cell technologies can convert stored energy in hydrogen to power with higher conversion efficiency compared to combustion engines, but the main barrier is the high production costs.

7.3.1 Reforming and Gasification Cases With CCUS

Earlier studies show that the autothermal reforming coupled with CCUS strategies has slightly better cost performance than other methods. The major cost drivers are the resources or feedstocks (e.g., natural gas or coal) in reforming processes and the capital costs in gasification processes. The latest study by the National Energy Technology Laboratory in 2022 provided all the details about the techno-economic and environmental impact assessment aspects, along with a block flow diagram of each hydrogen production pathway with CCUS (Table 7.3) [3]. The life cycle assessment results show that coal and biomass gasification with CCUS has the lowest GHG emissions (about -1 Ib CO_2 per Ib H_2), and natural gas autothermal reforming with CCUS has the highest GHG emissions (about 5.7 Ib CO_2 per Ib H_2) for a 100-year time horizon.

The levelized cost results of hydrogen production show that the natural gas autothermal reforming with CCUS and coal and biomass gasification with CCUS have the lowest ($1.51/kg) and highest ($3.44/kg) costs, respectively. The major cost drivers of natural gas autothermal reforming with CCUS include feedstock cost ($0.77/kg), operational cost ($0.47/kg), and capital cost ($0.26/kg). The major cost drivers of coal and biomass gasification with CCUS include capital cost ($1.46/kg), operational cost ($1.18/kg), and feedstock cost ($0.8/kg).

Table 7.3 An overview of techno-economic and environmental impact assessment of different hydrogen production technologies with CCUS.

| Technology | Major GHG emissions (lb CO_2 per lb H_2) | | | | | | Total estimated GHG emissions | Levelized cost ($/kg H_2) |
	Biomass emissions	CO_2 management emissions	Coal emissions	Grid power emissions	Natural gas emissions	Stack emissions[a]		
Coal gasification with CCUS	–	0.26	1.6	0.83	–	1.3	4.1	2.92
Coal/biomass gasification with CCUS	–4.0	0.32	1.1	–0.053	–	1.6	–1.0	3.44
Methane steam reforming with CCUS	–	0.15	–	1.2	2.9	0.38	4.6	1.54
Natural gas autothermal reforming with CCUS	–	0.14	–	2.4	2.7	0.51	5.7	1.51

a) Represent the uncontrolled discharge from the process.
Source: Adapted from Lewis et al. [3].

As explained in Chapters 2 and 6, there are several CCUS strategies for reducing GHG emissions during power and biofuel production. For example, chemical looping combustion is a CCUS approach that can be used for power generation from coal, and it can be mixed with other technologies (e.g., gasification) for fuel or gas production (e.g., biofuels and hydrogen) with net-zero CO_2 emissions. Chemical looping gasification of biomass is a new method with TRL 3 for high-quality syngas production that can be used for the downstream fuel synthesis process. The produced gases in the chemical looping gasification are mainly H_2, CO, CO_2, and H_2O, along with NH_3, H_2S, and tar components. Roshan Kumar et al. conducted a study and provided more results on the advantages and disadvantages of the chemical looping gasification of biomass process pathway [20].

Globally, several hydrogen production plants mixed with CCUS are operating or under development (Table 7.4). Most existing operating plants use steam methane reforming technology for hydrogen production at $2 per gallon, excluding delivery, storage, and dispensing. Currently, three operating plants (i.e., Air Products and Chemicals, Air Liquide, and Shell) use CCUS strategies to capture carbon (up to 60%). Most planned hydrogen production plants with carbon capturing will use natural gas autothermal reforming technology and improve CCUS techniques to capture carbon up to 98%. As discussed earlier, the main benefits of the natural gas autothermal reforming process include high energy efficiency (up to 80%) and high purity H_2 level that can address several industrial needs.

7.3.2 Water Electrolysis with Renewable Energy

Water electrolysis is a promising, sustainable hydrogen production pathway through low or high temperatures for direct water splitting, using various methods, such as thermochemical and photoelectrochemical, to generate hydrogen and oxygen from water. Thermally driven water splitting from diverse domestic and sustainable resources (e.g., concentrated solar heat or waste heat from nuclear reactors) can be used to produce hydrogen with low emissions. These pathways have the potential to address national priorities (e.g., energy security and economic opportunities) and environmental and sustainability benefits due to near-zero emissions when power is supplied from renewable and clean sources. However, the key challenges are hydrogen quality, scalability, control and safety, capital and operating costs, and technology requirements.

Thermochemical water splitting method uses high-temperature heat (500–2,000 °C) from various sources (e.g., solar, wind, nuclear power, or waste heat) for hydrogen and oxygen production from water. This technology has low to zero GHG emissions. Different solar-based thermochemical water-splitting processes have been investigated for hydrogen production, such as direct and hybrid cycles (Figure 7.11). The direct cycles are less complex, but they require

Table 7.4 Current operating and planned hydrogen production plants with CCUS.

Plant or Project	Technology	Capacity (million cubic feet per day)	Carbon Capture (%)	Status	Location	Study
Air Products and Chemicals, Inc	Steam methane reforming	200	60	Operating	United States	Air Products and Chemicals, Inc (APCI) Port Arthur Industrial Carbon Capture and Storage Project
Air Liquide	Steam methane reforming	45	60	Operating	France	Technology status of hydrogen production from fossil fuels with CCS
Shell	Steam methane reforming	191	50	Operating	Canada	Quest Carbon Capture and Storage (CCS) Project – Hydrogen Manufacturing Unit Material Balance
BEIS Hydrogen Supply Programme	Natural gas autothermal reforming	90	97	Planned	United Kingdom	HyNet Low Carbon Hydrogen Plant: Phase 1 Report for BEIS
H21	Natural gas autothermal reforming	2,900	94	Planned	United Kingdom	H21 North of England
Acron	Natural gas autothermal reforming	48	98	Planned	Scotland	Acorn Hydrogen Feasibility Study
Air Products	Natural gas autothermal reforming	623	95	Planned	Canada	Air Products Announces Multi-Billion Dollar Net-Zero Hydrogen Energy Complex in Edmonton, Alberta, Canada
Air Products	Natural gas autothermal reforming	750	95	Planned	United States	Louisiana Governor Edwards and Air Products Announce Landmark U.S. $4.5 Billion Blue Hydrogen Clean Energy Complex in Eastern Louisiana

Source: Adapted from Lewis et al. [3].

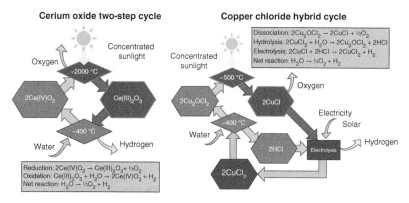

Figure 7.11 Example of hydrogen production through solar thermochemical water splitting. Source: U.S. Department of Energy (public domain) [6].

a higher temperature. The hybrid cycles can operate at low temperatures, but require extra equipment for on-site power inputs that increases the complexity.

The concentrated solar power can provide the high temperatures required for water splitting via direct thermal cycles (Figure 7.12). The main challenges are the efficiency and durability of the thermochemical reactor design (e.g., heat cycling) or reactant materials (e.g., cerium oxide or copper chloride). Other challenges are significant capital costs and a large area for efficient concentration, which is why this technology has low TRL and requires more investigation. Some of the strategies to address the existing barriers are: (i) improving reactant materials to optimize heat transfer and durability, (ii) reducing heat losses, (iii) improving

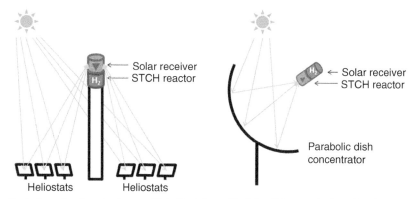

Figure 7.12 Schematic of concentrated solar power for hydrogen production via direct thermal cycles. Source: U.S. Department of Energy (public domain) [6].

the membrane materials to optimize conductivity and durability, (iv) improving efficiency by optimizing the electrolysis process, and (v) improving the thermochemical process by developing thermal and chemical storage systems. A recent National Renewable Energy Laboratory study provided more information about the techno-economic analysis of solar-based thermochemical hydrogen production [21].

Hydrogen production via direct seawater electrolysis is a new approach with TRL 3 that requires highly pure water with specific attributes and properties (e.g., resistivity, sodium, and chloride content) to achieve long-term stability [22]. The technology has been extensively studied recently since over 96% of Earth's water resource is seawater. However, the main challenges are: (i) the limited access to fresh water or highly pure water and (ii) high capital and operational costs of water purification. Additionally, seawater electrolysis can be used for producing oxygen or chloride for other industrial applications. There are several strategies to address the high capital and operational costs of hydrogen production through water electrolysis, including: (i) improving the materials and membrane electrode in terms of corrosion resistance, conductivity, and durability; (ii) optimizing the catalyst use and thermal integration; and (iii) improving the water conditioning efficiency.

Self-Check Questions

1. What are mature low-emission hydrogen production technologies?
2. What are the main demands of hydrogen?
3. What are the main methods for large-scale hydrogen production? And what are the key challenges?
4. What are the critical parameters for hydrogen production via thermochemical processes?
5. What is the most common method for hydrogen purification to achieve the required purity for the commercial scale?
6. What are the main steps of the pressure swing adsorption process?
7. What are the benefits and drawbacks of hydrogen production through thermochemical processes mixed with CCUS?
8. What is methane steam reforming?
9. What are the main components of methane steam reforming?
10. What are the key parameters of the methane steam reforming process?
11. What are the benefits and drawbacks of the methane steam reforming process?
12. What are the differences between carbon capturing before and after hydrogen separation in methane steam reforming?
13. What is natural gas autothermal reforming?
14. What are the key parameters for improving the output and efficiency of natural gas autothermal reforming?

15. What are the benefits and drawbacks of the autothermal reforming process?
16. What are the main differences between natural gas autothermal reforming and methane steam reforming?
17. What are the main feedstock types for hydrogen production?
18. What are the benefits and drawbacks of hydrogen production through biomass gasification?
19. What are the benefits and drawbacks of hydrogen production through methane pyrolysis?
20. What are the key components of electrolysis processes?
21. What are the benefits and drawbacks of hydrogen production through water electrolysis?
22. What are the key parameters affecting the process efficiency of alkaline electrolysis?
23. What are the benefits and drawbacks of alkaline electrolysis for hydrogen production?
24. What are the benefits and drawbacks of the proton-exchange membrane electrolysis process?
25. What are the key parameters affecting solid oxide electrolyzer performance?
26. What are the benefits and drawbacks of solid oxide electrolyzers?
27. What are the challenges and strategies for improving solar-based thermochemical hydrogen production?

References

1 U.S. DOE (2022). *Hydrogen Production.*

2 NETL (2022). *Guidelines/Handbook For The Design of Modular Gasification Systems.*

3 Lewis, E., McNaul, S., Jamieson, M. et al. (2022). *Comparison of Commercial, State-of-the-Art, Fossil-Based Hydrogen Production Technologies (No. DOE/NETL-2022/3241).* Pittsburgh, PA, Morgantown, WV, and Albany, OR (United States): National Energy Technology Laboratory (NETL).

4 Boretti, A. and Banik, B.K. (2021). Advances in hydrogen production from natural gas reforming. *Advanced Energy and Sustainability Research* 2 (11): 2100097.

5 Megía, P.J., Vizcaíno, A.J., Calles, J.A., and Carrero, A. (2021). Hydrogen production technologies: from fossil fuels toward renewable sources: a mini review. *Energy Fuels* 35 (20): 16403–16415.

6 U.S. DOE (2017). *Hydrogen Production Tech Team Roadmap.*

7 Ciambelli, P., Palma, V., Palo, E., and Iaquaniello, G. (2009). Natural gas autothermal reforming: an effective option for a sustainable distributed production of hydrogen. In: *Catalysis for Sustainable Energy Production*, 287–319. John Wiley & Sons, Ltd.

8 IEA (2021). *Net Zero by 2050: A Roadmap for the Global Energy Sector*.

9 Alptekin, F.M. and Celiktas, M.S. (2022). Review on catalytic biomass gasification for hydrogen production as a sustainable energy form and social, technological, economic, environmental, and political analysis of catalysts. *ACS Omega* 7 (29): 24918–24941.

10 Prajapati, B.K., Anand, A., Gautam, S., and Singh, P. (2022). Production of hydrogen-and methane-rich gas by stepped pyrolysis of biomass and its utilization in IC engines. *Clean Technologies and Environmental Policy* 24 (5): 1375–1388.

11 Lott, P., Mokashi, M.B., Müller, H. et al. (2023). Hydrogen production and carbon capture by gas-phase methane pyrolysis: a feasibility study. *ChemSusChem* 16 (6): e202201720.

12 Machhammer, O., Bode, A., and Hormuth, W. (2016). Financial and ecological evaluation of hydrogen production processes on large scale. *Chemical Engineering & Technology* 39 (6): 1185–1193.

13 NETL (2021). *Technologies for Hydrogen Production*. netl.doe.gov.

14 NETL (2021). *Hydrogen Production from Coal without Power Export*. netl.doe .gov.

15 Midilli, A., Kucuk, H., Topal, M.E. et al. (2021). A comprehensive review on hydrogen production from coal gasification: challenges and Opportunities. *International Journal of Hydrogen Energy* 46 (50): 25385–25412.

16 Pozio, A., Bozza, F., Lisi, N., and Mura, F. (2022). Electrophoretic deposition of cobalt oxide anodes for alkaline membrane water electrolyzer. *International Journal of Energy Research* 46 (2): 952–963.

17 International Energy Agency (2021). *Global Hydrogen Review 2021*. OECD Publishing.

18 Carmo, M. (2022). *Introduction to Liquid Alkaline Electrolysis*. U.S. Department of Energy/Office of Energy Efficiency and Renewable Energy/Hydrogen and Fuel Cell Technologies Office https://www.energy.gov/sites/default/files/2022-02/2-Intro-Liquid%20Alkaline%20Workshop.pdf.

19 Ahmad Kamaroddin, M.F., Sabli, N., Tuan Abdullah, T.A. et al. (2021). Membrane-based electrolysis for hydrogen production: a review. *Membranes (Basel)* 11 (11): 810.

20 Roshan Kumar, T., Mattisson, T., Rydén, M., and Stenberg, V. (2022). Process analysis of chemical looping gasification of biomass for Fischer–Tropsch crude production with net-negative CO2 emissions: part 1. *Energy & Fuels* 36 (17): 9687–9705.

21 Ma, Z., Davenport, P., and Saur, G. (2022). System and technoeconomic analysis of solar thermochemical hydrogen production. *Renewable Energy* 190: 294–308.

22 Khan, M.A., Al-Attas, T., Roy, S. et al. (2021). Seawater electrolysis for hydrogen production: a solution looking for a problem? *Energy Environ. Sci.* 14 (9): 4831–4839.

8

Hydrogen Storage

Hydrogen is one of the most promising energy sources for supporting the decarbonization of various sectors due to its high mass-based energy density (33–34 kWh/kg) compared to diesel or gasoline (around 12–14 kWh/kg). Hydrogen storage is crucial to advancing the hydrogen energy industry for various applications, such as power generation or transportation fuel. Hydrogen-based energy storage has gained high interest due to its high capacity and flexibility for medium- and long-term energy storage that cannot be achieved by other technologies. The existing, commercialized energy storage technologies are pumped hydro, compressed air, hydrogen-based, and batteries. Several hydrogen-based storage technologies exist (Figure 8.1), such as compressed gas in tanks for small volumes (up to 700 bar pressure) or underground storage for large-scale applications. The main technical parameters for hydrogen storage include: (i) weight, volume, and discharge rate; (ii) heat requirements; (iii) capital and operational costs; and (iv) storage capacity and recharging time [1].

The three primary forms of hydrogen storage are gas, liquid, and solid. Each form has its own benefits and drawbacks, which depend on its technical parameters and specific applications (e.g., portable or stationary). The most common form for hydrogen storage is the gas form in steel or composite tanks and requires high pressure (350–700 bar). Hydrogen gas cooled to near cryogenic temperatures (cryogas) is another way of storing hydrogen due to its high volumetric energy density. Cryogenic storage involves hydrogen storage as a supercritical fluid at low temperatures that can provide high storage capacity with high energy density, but it needs a cooling system. Hydrogen can be stored by converting to liquid form at extremely low temperatures (below −253 °C) for storing small volumes in insulated tanks to minimize heat transfer. Liquid storage offers higher energy density compared to gas storage, but it does require cryogenic infrastructure and loses around 40% of energy during the liquefaction process, which is why it has not been commercialized and is mainly used in high-technology applications, such as space travel.

Net-Zero and Low Carbon Solutions for the Energy Sector: A Guide to Decarbonization Technologies, First Edition. Amin Mirkouei.

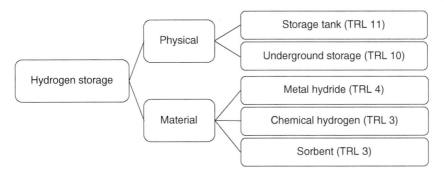

Figure 8.1 Mature hydrogen storage technologies.

Solid storage of hydrogen has been investigated by mixing it with solid materials that have hydrogen storage capacity, such as metal hydrides as absorbents or porous materials as adsorbents. In other words, hydrogen can be stored within (absorption) or on the surface (adsorption) of solids. This method stores hydrogen by chemically bonding with certain metals and alloys (e.g., magnesium and titanium) and forming metal hydrides that can offer reversible storage (store and release) under specific conditions. However, the main challenges for solid storage are the lack of sufficient reversibility and the complex hydrogen extraction process [2]. Most hydrogen storage systems also require a heat exchanger during charging and discharging. For heat requirements, waste heat from other sources (e.g., combustion or fuel cell processes) can be used to reduce energy consumption.

An overview of key solutions for hydrogen storage is presented in this chapter. Particularly, this chapter provides five technologies with high technology readiness levels, along with a case study about aboveground and underground hydrogen storage.

8.1   Physical-based Storage

Hydrogen-based energy storage can equilibrate supply and demand by storing excess energy (mainly renewable) when the supply exceeds demand. Also, it can be used for seasonal storage with a high capacity up to TWh due to its high energy density [2]. The key challenges are the proper discharge temperature and the ability to recharge the system quickly (a couple of minutes).

8.1.1   Storage Tanks (TRL 11)

Several hydrogen storage tanks (e.g., compressed gas, liquid hydrogen, and cryogenic hydrogen) have been used for different applications (Figure 8.2). The major

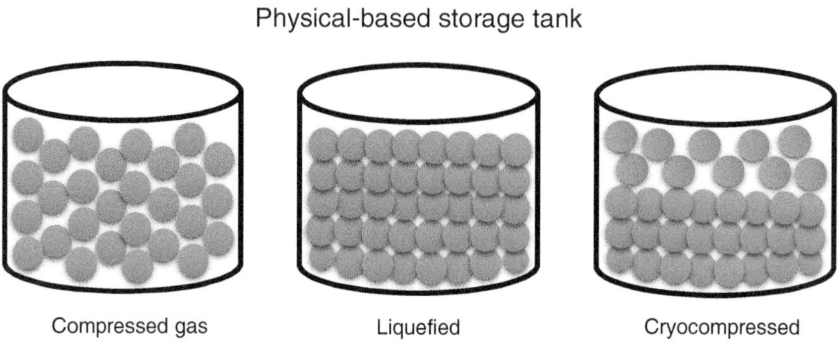

Physical-based storage tank

| Compressed gas | Liquefied | Cryocompressed |

Figure 8.2   Schematic of physical-based hydrogen storage forms.

parameters to consider for using hydrogen tanks include: (i) hydrogen properties (density, volume, and amount), (ii) material and design compatibility (resistance and external damage), (iii) temperature and pressure management for leakage and relief systems, (iv) safety measures (flammability and risks), and (v) storage methods (gas, liquid, or solid). Understanding these parameters helps develop efficient tanks with various application capabilities.

Compressed gas tanks have been used to store hydrogen gas under high pressure in different storage tanks, from steel, composite, or glass microspheres. The most common one is made of steel that can hold up to 10,000 psi (around 700 bar). Other compressed gas tanks are made of: (i) steel with a composite wrap, (ii) composite liner with a carbon fiber wrap, and (iii) plastic liner with a carbon fiber wrap for handling higher pressures. These tanks have been used mainly for stationary storage and fuel cell vehicles.

Lightweight composite tanks can endure high pressures (around 350–700 bar) and meet other safety and commercial aspects. Also, composite tanks can work without an internal heat exchanger and can be used for cryogas. These tanks are made of several layers, including an impact-resistant dome, high-density polymer liner, carbon-fiber-reinforced shell, and reinforced external protective shell. The main advantages are lightweight, corrosion resistance, size flexibility, gunfire and impact safety, abrasion resistance, and cost-competitiveness. However, the main drawbacks of composite tanks for hydrogen storage are: (i) large size and physical volume, which makes it difficult to adjust it with the available spaces; (ii) high cost (over $500 per kg); (iii) high-pressure requirements for compressing the gas; and (iv) safety challenges due to high pressure. These composite materials require further studies under various conditions (e.g., temperature and pressure) to assess their long-term effects.

Glass microspheres can be used for hydrogen storage, and the steps include charging, filling, and discharging. However, it requires high pressure and temperature (700 bar and 300 °C). After cooling down the microspheres to room temperature, they will be transferred to low-pressure tanks. Then, the microspheres will be heated up to 300 °C to release the hydrogen. The benefits of glass microspheres are: (i) a safer method for hydrogen storage in low-pressure compared to other methods, (ii) low container costs, and (iii) hydrogen storage density of 5.4 wt%. The key challenges with glass microspheres are: (i) low volumetric density, (ii) high pressure and temperature requirements for filling, and (iii) hydrogen leaks at ambient temperature.

Liquid hydrogen has a better storage density at relatively low pressures compared to compressed gas; however, it takes 30–40% extra energy to produce liquid hydrogen and requires super-insulated containers. Currently, the most common methods to store liquid hydrogen are cryogenic, borohydride ($NaBH_4$) solutions, and rechargeable organic liquids. The cryogenic process refers to cooling down hydrogen to liquid form with critical pressure of around 13 bar and temperature below −253 °C, as well as a volumetric and gravimetric storage density of approximately 70 kg/m^3 and 20 wt.%, respectively. Borohydride solutions can be used as a hydrogen storage medium, and the hydrolysis reaction is: $NaBH_4 + 2H_2O \rightarrow 4H_2 + NaBO_2$. In this method, the maximum theoretical gravimetric hydrogen storage density is around 11 wt.%, and the estimated cost is around 4.5–5 times more than the $NaBH_4$ cost. The benefits of borohydride solutions are onboard hydrogen generation, controllability, and safety, but the main drawback is $NaBH_4$ regeneration from $NaBO_2$. The rechargeable organic liquid method includes dehydrogenation, hydrogen refilling, and hydrogenation processes using various organic liquids such as methylcyclohexane (C_7H_{14}). The volumetric and gravimetric storage density of approximately 43 kg/m^3 and 6 wt.%. The main challenges of rechargeable organic liquid method are: (i) special handling of organic liquids due to their violent reactions, (ii) detailed safety and toxicity studies for fire and explosion hazards, and (iii) required infrastructure for dehydrogenation and hydrogenation processes. Hydrogen has almost three times more energy content (120 MJ/kg) than gasoline (44 MJ/kg) on a mass basis. However, liquid hydrogen has four times less density (8 MJ/L) compared to gasoline (32 MJ/L) on a volume basis (Figure 8.3).

8.1.2 Underground Storage (TRL 10)

Globally, most (around 75%) of the underground hydrogen storage is in depleted deposits (Figure 8.4). Hydrogen must be compressed for storing underground due to its low molecular weight, density, solubility in water, and dynamic viscosity. Earlier techno-economic studies show that underground hydrogen storage facilities are capital-intensive investments, and their economic viability depends on the required capital expenditure, especially for the construction of

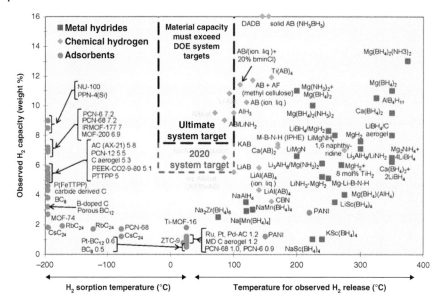

Figure 8.3 Energy density comparison of different energy sources and fuels on lower heating values. Source: U.S. Department of Energy (public domain).

deep geological storage sites. The benefits of underground hydrogen storage are: (i) high capacity (TWh for weeks or months) compared to aboveground storage sites with limited capacity (MWh for hours or days), (ii) lower storage costs, and (iii) high safety and less vulnerability to terrorist attacks. However, underground hydrogen storage involves gas leakage and losses due to faults, fractures, or biological reactions, which requires a leakage test with a maximum allowable leakage rate of $160\,m^3$/year [3]. One of the main reasons for hydrogen leakage is its low viscosity. The TRL of underground hydrogen storage methods varies from 4 to 10. Currently, salt cavern storage has the highest TRL (around 9–10), and other methods (e.g., deep aquifers or depleted oil and gas reservoirs) have a TRL of around 3–4 due to many barriers, such as storage efficiency, geology challenges, safety requirements, and legal limitations.

Salt caverns are mainly cylindrical pits in thick underground salt deposits and depend on geological conditions (e.g., salt mechanical properties, tightness, and resistance to chemical reactions). Salt cavern storage has been used for large-scale hydrogen storage for medium to long periods, enabling seasonal hydrogen gas storage with up to 10 times injection and extraction per year. This method has gained high interest due to its stability, ease of management, and high volume capacity of up to 1 million m^3 at a maximum pressure of 200 bar [4]. However, this technique

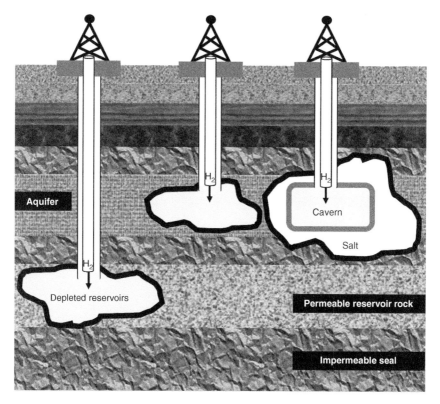

Figure 8.4 Schematic of underground hydrogen storage techniques.

has several technical challenges, such as the installation, transfer capacity, and tightness of boreholes, as well as location and environmental limitations.

Depleted oil and gas reservoirs, or deep aquifers, have been used for large-scale hydrogen storage for medium to long periods, and their capacity depends on geological conditions. Depleted oil and gas reservoirs are geological traps with impermeable layers, supported by aquifers. These reservoirs are known options for hydrogen storage due to their reliable tightness and geological structures. Aquifers are permeable, porous media that can store gas or water in their pore space. The main parameters for using aquifers for hydrogen storage are rock structure and properties and impermeable layers for avoiding gas migration. This method enables seasonal hydrogen gas storage with a few annual injection and withdrawal cycles. Due to greater geological distribution, this method is much more flexible for large-scale hydrogen storage than salt caverns. The capacity of aquifers can be determined by their geological structure (e.g., reservoir rocks, size, and thickness). Some barriers to this method are: (i) uneven distributions and complexity, (ii) lack of reliable storage capacity, and (iii) high costs of finding reliable storage sites.

Understanding the various interactions (e.g., geochemical, microbiological, and geomechanical) during underground hydrogen storage are critical and necessary to address some barriers (e.g., reducing leakage and safety concerns and improving efficient industrial-scale underground hydrogen storage). The geochemical interactions depend on the chemical and mineral composition of reservoir rocks and fluids, reservoir pressure and temperature, porous rock structure, and mineralogical changes. The microbiological interactions highly depend on the presence of microorganisms in underground storage sites, particularly hydrogen-consuming microorganisms (hydrogenotrophic) that can lead to loss of gas. The geomechanical interactions mainly depend on cyclic injection and collection of hydrogen that can affect rock formation and stability and lead to cracks and fractures due to stress fluctuations. The scale of interactions varies between different storage methods (e.g., depleted hydrocarbon reservoirs, deep aquifers, and salt caverns). The salt caverns method has fewer interactions compared to other methods.

8.2 Material-based Storage

Material-based storage is another approach for storing hydrogen, mainly in the solid form, using different materials (e.g., metal hydrides, chemical hydrogen, and adsorbents) to improve critical factors, such as volumetric and gravimetric capacities, required storage pressure and temperature, and cost (Table 8.1). Currently, material-based hydrogen storage studies focus on advancing the storage materials, including chemical composition and distribution and material structure (microscopic and macroscopic).

Table 8.1 Material-based hydrogen storage attributes.

Attributes	Unit	Metal hydrides	Adsorbents	Chemical Hydrogen
Volumetric density	kWh/L	0.4	0.7	1.3
	kg H_2/L	0.012	0.021	0.040
Gravimetric density	kWh/kg	0.4	1.3	1.5
	kg H_2/kg	0.012	0.038	0.046
Cost	$/kWh	43	15	17
	$/kg H_2	1430	490	550

Source: Adapted from U.S. DOE [5].

Figure 8.5 Gravimetric capacity of many unique hydrogen storage materials: Source: U.S. Department of Energy, Hydrogen and Fuel Cell Technologies Office (public domain) [5].

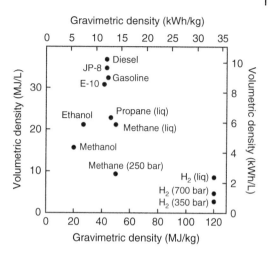

Over the past two decades, several studies have investigated over 400 materials for hydrogen storage (Figure 8.5). The major performance factors for material-based hydrogen storage are kinetics, capacity, thermodynamics, and cycle life, as well as thermal and mechanical properties. Earlier studies provided the recommended best practices, methodologies, and protocols for measuring hydrogen storage material properties [6].

8.2.1 Metal Hydride Materials (TRL 4)

Metal hydride hydrogen storage is the most common method due to various applications, such as electrochemical cycling [7], thermal storage [8], and heat pumps [9]. The latest studies focus on complex hydrides, consisting of alkaline earth elements that can be ionically bonded to a complex anion (e.g., Ni, Fe, Al, and B) [10]. Complex hydrides can offer higher hydrogen density (gravimetric and volumetric) materials. The desorption reaction during the hydrogen release can be endothermic. Further investigation is needed to improve reaction thermodynamics and long-term cycling, along with hydrogen adsorption and desorption kinetics.

8.2.2 Chemical Hydrogen Materials (TRL 3)

Chemical hydrogen storage method uses materials or compounds with high hydrogen density that can bind with hydrogen in solid or liquid form. Hydrogen release from chemical compounds is mainly exothermic. Currently, the main challenge of this method for large-scale transportation is the rehydrogenation process of the spent dehydrogenated materials. Currently, off-board rehydrogenation

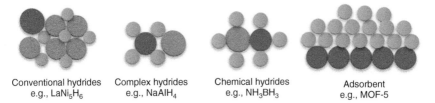

Conventional hydrides
e.g., LaNi$_5$H$_6$

Complex hydrides
e.g., NaAlH$_4$

Chemical hydrides
e.g., NH$_3$BH$_3$

Adsorbent
e.g., MOF-5

Figure 8.6 Schematic of material-based hydrogen storage methods. Source: Adapted from U.S. DOE [15].

can be used, which greatly increases operational costs [11]. The dehydrogenation reaction pathways are with water [12] or heating the chemical compounds with high hydrogen capacities, such as ammonia borane (NH$_3$BH$_3$) and aluminum hydride (AIH$_3$) [13]. Other challenges include handling the dehydrogenated spent materials and remaining them in the liquid phase.

8.2.3 Sorbent Materials (TRL 3)

Sorbent hydrogen storage method uses adsorbents as a storage medium. The benefit is that dihydrogen retains its form throughout the adsorption and desorption cycles with low activation energy; however, the adsorbents have weak adsorption enthalpies compared to metal hydrides or chemical hydrogen storage materials [14]. The adsorption enthalpy range for hydrogen adsorbents can be around 5–10 kJ/mole. Various materials with high gravimetric density have been investigated as adsorbents for hydrogen storage, such as polymers or activated carbons. The main two parameters for high gravimetric adsorption are surface area and micropore volume. Latest studies focus on improving hydrogen volumetric capacity and dihydrogen binding energies, as well as the material pore size and surface area. Figure 8.6 presents the mature material-based hydrogen storage methods.

8.3 Case Study: Aboveground and Underground Hydrogen Storage

8.3.1 Aboveground Hydrogen Storage

Currently, the dominant hydrogen storage system for vehicles is compressed hydrogen at high pressures (e.g., 350 and 700 bar). Recent studies have investigated different fuel cell technologies and hydrogen storage systems in vehicles to provide a driving range of around 300 miles and meet cost, safety, and performance requirements (e.g., capacity, refiling quickly and easily, discharge kinetics, operability, and durability) [15]. Earlier studies in 2010 showed that the hydrogen storage capacities for 350 and 700 bar are 0.017 and 0.025 kg hydrogen/L, and the costs are $19 and $16 per kWh, respectively. The major challenge is to store high-density hydrogen for stationary or portable applications. Light-duty hydrogen fuel cell electric vehicles can meet the requirements using large-volume, high-pressure compressed gas composite storage tanks. Based on the hydrogen energy content and density, light-duty vehicles need hydrogen storage with 5–13 kg capacities to meet the driving range and commercialization aspects (Table 8.2).

Composite tanks for hydrogen storage have several benefits: they do not require heat exchanger; they have been extensively tested, and they comply with available codes and standards. However, the drawbacks include requiring a high-pressure compressor and high storage cost (over $500/kg of hydrogen). Liquid hydrogen storage, using the cryogenic process, has been demonstrated in commercial vehicles (e.g., BMW) and can be used in airplanes due to its high storage density. This method requires more efficient liquefaction technologies, advanced insulated containers, and reduced costs. Also, a few commercial companies (e.g., Millennium Cell in the United States and MERIT in Japan) have used borohydride solutions for liquid hydrogen storage. Comparing different liquid hydrogen storage methods shows that they all require a couple of factors, such as safe and well-organized handling, transportation, and infrastructure due to toxic chemicals or extreme temperatures [17]. Storing hydrogen in solid materials has several benefits, such as a safe and efficient approach for portable

Table 8.2 Hydrogen storage for light-duty fuel cell vehicles.

Storage Parameter	2020	2025 (target)	Unit
Energy content	1.5	1.8	kWh/kg
Energy density	1.0	1.3	kWh/L
Storage cost	10	9	$/kWh
	333	300	$/kg
Fuel cost	4.0	4.0	$/gge
Operating temperature	−40/60	−40/60	°C
Delivery temperature (Min/Max)	−40/85	−40/85	°C
Delivery pressure (Min/Max)	5/12	5/12	Bar
Onboard efficiency	90	90	%
Charging time	3–5	3–5	Min

Source: Adapted from U.S. DOE [16].

and stationary applications. The suitable materials for solid storage include: (i) carbon-based materials (e.g., activated charcoal, nanotubes, and graphite nanofibers) or high-surface area materials (e.g., zeolites, clathrate hydrates, and metal oxides), (ii) water-reactive chemical hydrides (e.g., NaH, CaH_2, MgH_2) and thermal chemical hydrides (e.g., AlH_3 and NH_3BH_3), and (iii) rechargeable hydrides (e.g., alloys and nanocrystalline). The benefits of solid storage compared to gas or liquid storage are lower volume and pressure and higher hydrogen purity. Table 8.3 compares the current status, challenges, and best options of different hydrogen storage forms.

Several projects and teams in the United States and other countries are working on the scientific gaps blocking the advancement of hydrogen storage systems, such as the Hydrogen Materials Advanced Research Consortium (HyMARC), including researchers at the Sandia National Laboratories, National Renewable Energy Laboratory, Pacific Northwest National Laboratory, Lawrence Livermore National Laboratory, Lawrence Berkeley National Laboratory, SLAC National Accelerator Laboratory, and National Institute of Standards and Technology [18]. HyMARC mainly focuses on thermodynamic and kinetic limitations (e.g., mass transport and surface chemistry), innovative synthetic approaches and protocols, high-resolution in situ characterization methods, material research and development, and computational modeling and validation to improve onboard vehicular hydrogen storage.

Table 8.3 Hydrogen storage attributes in different forms.

Form	Status	Challenges	High TRL Option
Gas	Commercially available	Cost, safety, required compression energy, volume reduction, and fracture mechanics	Carbon-fiber-reinforced composite tanks (up to 10 wt.% hydrogen at 700 bar) for near-term, small-scale distributed systems
Liquid	Commercially available	Cost, safety, required liquefaction energy, and dormant boil-off	Cryogenic insulated containers (20 wt.% hydrogen at −253 °C and 1 bar) for near to medium-term, large-scale centralized systems
Solid	Early deployment	Cost, compatibility, pressure and heat management, weight, and optimization	Metal hydrides (8 wt.% hydrogen at 60 bar) for long term in distributed systems

Source: Adapted from International Energy Agency [1].

8.3.2 Underground Hydrogen Storage

Underground hydrogen storage has a high potential for the low-carbon economy due to its quick implementation. We have limited underground hydrogen storage experience compared to underground natural gas storage (e.g., salt caverns or porous rocks). Recent studies have shown that there are over 660 underground natural gas storage sites, mainly in depleted hydrocarbon reservoirs (72%), salt caverns (15%), and deep aquifers (11%) [3]. Several underground hydrogen storage projects have been investigated in the United States and Europe (e.g., HyUnder [19] and Hystories [20]) to assess the geological and geographical factors, the economic viability of renewable power conversion to hydrogen and underground storage, as well as other large-scale hydrogen transportation, implementation, and legal barriers.

Currently, several facilities use underground hydrogen storage in the United States (located in Clemens, Moss Bluff, and Spindletop) and the United Kingdom (located in Teesside and Yorkshire) [4]. The U.S. projects use 95% hydrogen content, a storage volume of 560,000–580,000 m^3, a storage depth of 820–930 m, pressure around 55–152 bar, and stored energy of 90–120 GWh. Also, other underground hydrogen storage facilities in Germany, France, Czech Republic, Argentina, and Austria use porous rocks sites (e.g., aquifers and depleted gas reservoirs) with various hydrogen content (10–90%), storage depths (400–1,000 m), pressure (10–90 bar), and temperature (34–50 °C).

Finding cost-effective underground hydrogen storage sites requires expensive studies, including careful selection and assessment of various factors, such as geological structure, hydrogen interactions in storage sites, and reliable estimates of hydrogen storage capacity. Most of the low-TRL methods are not profitable, and the payback period is over 15 years, mainly due to the low demand for hydrogen fuel. Transporting hydrogen with long-distance pipelines costs 3–5 times less than other methods, such as by ship.

Self-Check Questions

1. What are mature hydrogen-based storage technologies?
2. What are the main technical parameters for hydrogen storage?
3. What are the main forms of hydrogen storage?
4. What does cryogenic hydrogen storage mean? And what are the benefits?
5. What are the main challenges for solid hydrogen storage?
6. What are the key challenges of physical-based hydrogen storage?
7. What are the major parameters for using hydrogen storage tanks?
8. What are the main physical-based hydrogen storage forms?
9. What are the main advantages of lightweight composite hydrogen storage tanks?

10. What are the main drawbacks of composite tanks for hydrogen storage?
11. What are the benefits and drawbacks of hydrogen storage with glass microspheres?
12. What are the benefits and drawbacks of liquid hydrogen storage compared to gas storage?
13. What are the most common liquid hydrogen storage methods?
14. What are the mature underground storage methods?
15. What are the benefits and drawbacks of underground hydrogen storage?
16. What is the salt cavern method for hydrogen storage?
17. What is a depleted oil and gas reservoir for hydrogen storage?
18. What is the aquifer method for hydrogen storage? And what are the pros and cons of this method?
19. What is material-based hydrogen storage?
20. What are the mature material-based hydrogen storage methods?
21. What are the suitable materials for solid hydrogen storage?
22. What are the benefits of solid storage compared to gas or liquid storage?

References

1 International Energy Agency (2006). *Hydrogen Production and Storage.*
2 Yue, M., Lambert, H., Pahon, E. et al. (2021). Hydrogen energy systems: a critical review of technologies, applications, trends and challenges. *Renewable and Sustainable Energy Reviews* 146: 111180.
3 Tarkowski, R. and Uliasz-Misiak, B. (2022). Towards underground hydrogen storage: a review of barriers. *Renewable and Sustainable Energy Reviews* 162: 112451.
4 Zivar, D., Kumar, S., and Foroozesh, J. (2021). Underground hydrogen storage: a comprehensive review. *International Journal of Hydrogen Energy* 46 (45): 23436–23462.
5 U.S. DOE (2022). *Materials-Based Hydrogen Storage.* Energy.gov.
6 U.S. DOE (2014). *Recommended Best Practices for the Characterization of Storage Properties of Hydrogen Storage Materials.* Energy.gov.
7 Young, K. and Nei, J. (2013). The current status of hydrogen storage alloy development for electrochemical applications. *Materials* 6 (10): 4574–4608.
8 Felderhoff, M. and Bogdanović, B. (2009). High temperature metal hydrides as heat storage materials for solar and related applications. *International Journal of Molecular Sciences* 10 (1): 325–344.
9 Lototskyy, M.V., Yartys, V.A., Pollet, B.G., and Bowman, R.C. Jr., (2014). Metal hydride hydrogen compressors: a review. *International Journal of Hydrogen Energy* 39 (11): 5818–5851.

10 U.S. DOE (2022). *Metal Hydride Storage Materials*. Energy.gov.

11 U.S. DOE (2022). *Chemical Hydrogen Storage Materials*. Energy.gov.

12 Kojima, Y., Kawai, Y., Kimbara, M. et al. (2004). Hydrogen generation by hydrolysis reaction of lithium borohydride. *International Journal of Hydrogen Energy* 29 (12): 1213–1217.

13 Gutowska, A., Li, L., Shin, Y. et al. (2005). Nanoscaffold mediates hydrogen release and the reactivity of ammonia borane. *Angewandte Chemie* 117 (23): 3644–3648.

14 U.S. DOE (2022). *Sorbent Storage Materials*. Energy.gov.

15 U.S. DOE (2022). *Hydrogen Storage*. Energy.gov.

16 U.S. DOE (2022). *DOE Technical Targets for Onboard Hydrogen Storage for Light-Duty Vehicles*. Energy.gov.

17 Zhang, T., Uratani, J., Huang, Y. et al. (2023). Hydrogen liquefaction and storage: recent progress and perspectives. *Renewable and Sustainable Energy Reviews* 176: 113204.

18 U.S. DOE (2022). *HyMARC: Hydrogen Materials Advanced Research Consortium*. Energy.gov.

19 HyUnder (2012). *Assessing the Potential, Actors and Business Models of Large Scale Underground Hydrogen Storage in Europe*.

20 Hystories (2023). *HYdrogen STORage in European Subsurface*.

Appendix 1

Opportunities to Study or Work with the Author

Dr. Mirkouei spent his education and career learning the science of different technologies, especially decarbonization technologies, to address existing regional and national challenges, such as air and water pollution, soil contamination, and biodiversity loss, along with energy and food security. His experience serves as the foundation for all his research and development, as well as teaching and disseminating the results and outcomes. His expertise and experiences resulted in the development of several courses in Industrial Technology, Technology Management programs, Mechanical Engineering, Biological Engineering, and Environmental Science programs at the University of Idaho. His knowledge and courses are designed to help students, engineers, scientists, technologists, managers, and politicians accelerate their learning and increase their business success.

In brief, Dr. Mirkouei has been working and planning to undertake more innovative research and technology development in the following fields:

- *Chemical and biochemical compounds production from biomass feedstocks or urban wastes*: The overarching goal of this multi-disciplinary Thrust is to advance the existing conversion processes for engineered materials and renewable material production (e.g., hydrocarbon fuels and engineered biochar) from various waste streams (e.g., biomass feedstocks, plastic wastes, e-wastes, and animal manure) through integrating the new mechanical inventions with growing cyber-based control and optimization initiatives. For example, biofuel conversion process represents the most substantial portion of the total biofuel cost (over 60%) based on recent techno-economic studies. Further investigation is needed not only in empirical work for specific applications, but also for the creation of conceptual cyber-infrastructures to advance the operations and increase the technical growth. To accomplish the goal of this Thrust, we are pursuing two research tasks: (i) modernizing the physical-based interface of the waste conversion process using programmable logic controllers, and developing new upgrading technologies (e.g., physicochemical pathways);

Net-Zero and Low Carbon Solutions for the Energy Sector: A Guide to Decarbonization Technologies, First Edition. Amin Mirkouei.

and (ii) constructing an adaptive cyber-based platform for data-driven decision-making, including intelligent control and optimization modules.

- *Precision agriculture, soil–plant health improvement, and food processing*: According to the U.S. Department of Agriculture (USDA), Idaho ranks in the top ten in the nation in nearly 30 of 168 agriculture commodities. The state ranks number one in the nation in the production of potatoes (34%) and Austrian winter peas. Idaho grows 70% of the hybrid temperate sweet corn seed produced in the world. For alfalfa, hay, hops, mint, and sugar beets, Idaho is number two in production for the nation. For fresh prunes and plums, Idaho is number three in production for the nation. Idaho ranks 1st in the nation for production of food-sized trout with 46%. Therefore, technology breakthroughs are essential for addressing agricultural challenges, e.g., crop quality and productivity. The main goal of this Thrust is to promote Idaho crop production sustainability through the use of renewable materials, slow-release fertilizers, and soil conditioners, as well as advanced cyber-based systems (e.g., data analytics and communication systems), and intelligent mechanical inventions (e.g., remote sensors) for fertigation (injection of chemicals into an irrigation system), which is one of the major cost-drivers in farming. Integrated analysis of agricultural and soil ecosystems is growing with improvements in technologies, standards, and data-driven decision-making. To accomplish the goal of this Thrust, we aim to develop renewable products and sustainable technologies, as well as construct and empirically verify a self-adaptive cyber-physical system for real-time sensing, control, and optimization of water-fertilizer use efficiency and soil–plant health improvement, using slow-release fertilizers and soil conditioners.

- *Sustainable aquaculture and water treatment*: Idaho's aquaculture industry ranks as the 3rd largest food-animal industry in the state and is the nation's largest commercial producer of rainbow trout. Also, Idaho ranks 3rd in the nation for milk production. Annually produced waste (both solid and liquid pollutants) from the food-animal and aquaculture industries has negative impacts on surface water and groundwater that can pose eutrophication and other environmental problems. For example, the wastewater discharged from fish hatcheries can contain uneaten fish food, fish feces, nutrients (especially phosphorus), algae, parasites and pathogens, drugs, and other chemicals. The overarching aim of this Thrust is to promote Idaho fish production sustainability through the use of renewable materials, such as carbonized biomass or engineered biochar, bridge the gap between knowledge discovery and technology (e.g., new tools and machines) implementation in manufacturing, and improve the performance of the growing aquaculture sector, by meeting often-conflicting sustainability objectives (e.g., economic, environmental, and social) through different sustainability approaches (e.g., life cycle assessment, energy monitoring and analysis, process energy optimization strategies, and social assessment methods). To accomplish the goal of this Thrust and address sustainability challenges in Idaho, we aim to conduct life cycle assessment (LCA) and input–output analysis to find the major pollution sources and find practical solutions to reduce the environmental emissions, such as blended wood-based biochar that has application in water-nutrient recycling at fish farms.

- *Rare earth elements exploration and recovery from electronic wastes and ore in Idaho*: According to the U.S. Geological Survey (USGS) and Idaho Geological Survey (IGS), Idaho is ranked as one of the most favorable mining jurisdictions that has compiled a national inventory of critical mineral resources, with two of the most promising prospects. The overarching goals of this Thrust are to expand Idaho-sourced rare earth elements (REEs) research and development in central Idaho and identify, coordinate, and develop research expertise. To accomplish the goal of this Thrust, we aim to (i) identify the areas that have enhanced concentrations of REEs in collaboration with IGS and our industry partner, Idaho Strategic Resources (IDR), (ii) develop processing and extraction practices for mixed rare earth oxide (REO) production in collaboration with Idaho National Laboratory (INL) and Critical Materials Institute (CMI), and (iii) conduct techno-economic analysis (TEA) and environmental impacts assessment or LCA for both upstream and midstream segments. Our current collaboration with IGS and IDR is beneficial to UI by contributing to REEs exploration projects (Diamond Creek and Roberts) and planning initial mapping, sampling, and drilling 10–20 holes over the next 12 months. IDR's vision is to expand Idaho-sourced REEs exploration in central Idaho and address national priorities, such as reducing U.S. reliance on foreign suppliers and promoting domestic job creation. IDR partnership with my research group, along with support from Idaho IGEM-Commerce, would provide critical support, enable the commercialization of IDR's current products, and help establish additional intellectual property (IP) on new material configurations and product lines.

Here is the list of existing courses and programs if you are willing to learn more about decarbonization solutions:

- *Sustainable Food-Energy-Water Systems (BE/ME 524)*: This course covers sustainability analysis, life cycle assessment, and applications of sustainability across design and manufacturing processes, as well as food-energy-water systems, which establishes the concept of sustainability, and sustainable engineering. This course introduces the intersection of sustainability and food-energy-water systems through sustainable development, sustainability principles, and environmental analysis.

- *Industrial Sustainability Analysis (INDT419/TM519)*: This course covers tools and information to help the students achieve organizational results and increase or build their skills and knowledge in sustainability and life cycle assessment (i.e., techno-economic and socio-environmental studies), as well as applications of sustainability in manufacturing processes and systems. This course focuses on the concepts, analysis methods, life cycle assessment tools, and metrics through direct involvement in interdisciplinary consulting projects.

- *Sustainable Manufacturing*: This online class is a week-long program that covers basic sustainability concepts and principles, along with critical skills and tools required to enhance sustainability benefits across manufacturing processes and systems. Dr. Mirkouei will also explain different companies and projects and answer the most frequent questions, such as how to improve process efficiency and productivity, and subsequently reduce environmental impacts and costs. This class is exceptionally powerful for business owners and startup companies that want to increase their revenue and growth with actual plans by avoiding common mistakes, and assessing techno-economic and environmental impact aspects. Participants will receive a comprehensive educational package, including high-TRL solutions and real-world project libraries.

- *Private One-On-One Consulting*: If you need a private advisor and guide to assess different decarbonization technologies and solutions, this program provides the basic science, key principles, and personal consulting, including directions and feedbacks, for 6–12 times per year (each meeting three to four hours). This program is exceptionally powerful for decision-makers to leverage their resources and results. You can contact Dr. Mirkouei at a.mirkouei@gmail.com.

Appendix 2

Author's Publications

Books, Magazines, and Contributions to Other Volumes

1 Mirkouei, A. (2021). Biochar from cow manure could be key to sustainable agriculture. *Forbes*. https://www.forbes.com/sites/aminmirkouei/.
2 Mirkouei, A. (2021). A renewable solution for polluted waters: biochar explained. *Forbes*. https://www.forbes.com/sites/aminmirkouei/.
3 Mirkouei, A. (2021). A sustainable way to mine rare earth elements from old tech devices: agromining explained. *Forbes*. https://www.forbes.com/sites/aminmirkouei/.
4 Mirkouei, A. (2019). Cyber-physical real-time monitoring and control: a case study of bioenergy production. In: *Emerging Frontiers in Industrial and Systems Engineering: Success through Collaboration*, 1e (ed. Nembhard, Cudney, and Coperich), Taylor & Francis. **(Book of the month)**

Refereed Journal Publications

1 Kerner, P., Struhs, E., Mirkouei, A. et al. (2023). Microbial Responses to Biochar Soil Amendment and Influential Factors: A Three-level Meta-analysis. *Environmental Science & Technology* https://doi.org/10.1021/acs.est.3c04201 **(IF: 11.4)**.
2 Bare, R., Struhs, E., Mirkouei, A. et al. (2023). Controlling Eutrophication of Aquaculture Production Water Using Biochar: Correlation of Molecular Composition with Adsorption Characteristics as Revealed by FT-ICR Mass Spectrometry. *Processes* 11 (10): 2883. https://doi.org/10.3390/pr11102883 **(IF: 3.5)**.
3 Albor, G., Mirkouei, A., McDonald, A.G. et al. (2023). Fixed bed batch slow pyrolysis process for polystyrene waste recycling. *Processes* 11 (4): 1126. https://doi.org/10.3390/pr11041126 **(IF: 3.3, Selected for the cover page)**.

Net-Zero and Low Carbon Solutions for the Energy Sector: A Guide to Decarbonization Technologies, First Edition. Amin Mirkouei.
© 2024 John Wiley & Sons, Inc. Published 2024 by John Wiley & Sons, Inc.

4 Bare, R., Struhs, E., Mirkouei, A. et al. (2023). Engineered, porous-structured biomaterials for removing harmful nutrients from downstream water of aquaculture facilities. *Processes* 11 (4): 1029. https://doi.org/10.3390/pr11041029 **(IF: 3.3, Featured paper)**.

5 Brown, R., Mirkouei, A., Reed, D., and Thompson, V. (2023). Current nature-based biological practices for rare earth elements extraction and recovery: Bioleaching and biosorption. *Renewable and Sustainable Energy Reviews* 173: https://doi.org/10.1016/j.rser.2022.1130 **(IF: 16.8)**.

6 Struhs, E., Sotoudehnia, F., Mirkouei, A. et al. (2022). Effect of feedstocks and free-fall pyrolysis on bio-oil and biochar attributes. *Journal of Analytical and Applied Pyrolysis* https://doi.org/10.1016/j.jaap.2022.105616 **(IF: 6.4)**.

7 Struhs, E., Mirkouei, A., Ramirez-Corredores, M.M. et al. (2021). Overview and technology opportunities for thermochemically-produced bio-blendstocks. *Journal of Environmental Chemical Engineering* https://doi.org/10.1016/j.jece.2021.106255 **(IF: 5.9)**.

8 Butte, S., Vakanski, A., Duellman, K., and Mirkouei, A. (2021). Potato crop stress identification in aerial images using deep learning-based object detection. *Agronomy Journal* https://doi.org/10.1002/agj2.20841 **(IF: 2.2)**.

9 Thompson, M., Mohajeri, A., and Mirkouei, A. (2021). Comparison of pyrolysis and hydrolysis processes for furfural production from sugar beet pulp: A case study in Southern Idaho, USA. *Journal of Cleaner Production* 127695: https://doi.org/10.1016/j.jclepro.2021.127695 **(IF: 9.2)**.

10 Opare, E.O., Struhs, E., and Mirkouei, A. (2021). A comparative state-of-technology review for rare earth element separation. *Renewable and Sustainable Energy Reviews* 143: 110917. https://doi.org/10.1016/j.rser.2021.110917 **(IF: 14.9)**.

11 Struhs, E., Hansen, S., Mirkouei, A. et al. (2021). Ultrasonic-assisted catalytic transfer hydrogenation for upgrading pyrolysis-oil. *Ultrasonics Sonochemistry* 73: 105502. https://doi.org/10.1016/j.ultsonch.2021.105502 **(IF: 6.5)**.

12 Struhs, E., Mirkouei, A., You, Y., and Mohajeri, A. (2020). Techno-economic and environmental assessments for nutrient-rich biochar production from cattle manure: A case study in Idaho, USA. *Applied Energy* 279: 115782. https://doi.org/10.1016/j.apenergy.2020.115782 **(IF: 9.7)**.

13 Mirkouei, A. (2020). A cyber-physical analyzer system for precision agriculture 4.0. *Journal of Environmental Science: Current Research* 3: 016. https://doi.org/10.24966/ESCR-5020/100016.

14 Hansen, S., Mirkouei, A., and Diaz, L. (2020). A comprehensive state-of-technology review for upgrading bio-oil to renewable or blended hydrocarbon fuels. *Renewable and Sustainable Energy Reviews* 118: 109548. https://doi.org/10.1016/j.rser.2019.109548 **(IF: 12.1)**.

15 Hersh, B., Mirkouei, A., Sessions, J. et al. (2019). A review and future directions on enhancing sustainability benefits across food-energy-water systems: The potential role of biochar-derived products. *AIMS Environmental Science* 6 (5): 379–416. https://doi.org/10.3934/environsci.2019.5.379.

16 Mirkouei, A., Haapala, K.R., Sessions, J., and Murthy, G.S. (2017). A mixed biomass-based energy supply chain for enhancing economic and environmental sustainability benefits: A multi-criteria decision making framework. *Applied Energy* https://doi.org/10.1016/j.apenergy.2017.09.001 **(IF: 8.4)**.

17 Mirkouei, A., Haapala, K.R., Sessions, J., and Murthy, G.S. (2017). A review and future directions in techno-economic modeling and optimization of upstream forest biomass to bio-oil supply chains. *Renewable and Sustainable Energy Reviews* RSER6200: https://doi.org/10.1016/j.rser.2016.08.053 **(IF: 10.5)**.

18 Mirkouei, A., Mirzaie, P., Haapala, K.R. et al. (2016). Reducing the cost and environmental impact of integrated fixed and mobile bio-oil refinery supply chains. *Journal of Cleaner Production* 113 (2016): 495–507. https://doi.org/10.1016/j.jclepro.2015.11.023 **(IF: 6.3)**.

19 Davarpanah, M.A., Mirkouei, A., Yu, X. et al. (2015). Effects of incremental depth and tool rotation on failure modes and microstructural properties in single point incremental forming of polymers. *Journal of Materials Processing Technology* https://doi.org/10.1016/j.jmatprotec.2015.03.014 **(IF: 4.1)**.

Peer-Reviewed Conference Publications

1 Zirker, D., Mirkouei, A., Duellman, K.M. et al. (2023). Phytomining of Idaho-sourced rare earth elements. *152nd TMS (The Minerals, Metals & Materials Society) Annual Meeting and Exhibition*, San Diego CA (19–22 March 2023) (poster presentation).

2 Jamil, H.M., Li, L., and Mirkouei, A. (2022). MatFlow: a system for knowledge-based novel materials design using machine learning. *IEEE International Conference on Big Data*. https://ieeexplore.ieee.org/document/10020246.

3 Albor, G., Mirkouei, A., and Struhs, E. (2022). Mixed plastic waste conversion to value-added products: sustainability assessment and a case study in Idaho. *Proceedings of the ASME IDETC/CIE: 27th Design for Manufacturing and the Life Cycle Conference*. https://doi.org/10.1115/DETC2022-89199 **(Received the Best Paper Award)**

4 Walters, J., Mirkouei, A., and Makrakis, G.M. (2022). A quantitative approach and an open-source tool for social impacts assessment. *Proceedings of the ASME IDETC/CIE: 27th Design for Manufacturing and the Life Cycle Conference*. https://doi.org/10.1115/DETC2022-89196.

5 Ohene Opare, E. and Mirkouei, A. (2021). Environmental and economic assessment of a portable E-waste recycling and rare earth elements recovery process. *Proceedings of ASME IDETC/CIE: 26th Design for Manufacturing and the Life Cycle Conference,* https://doi.org/10.1115/DETC2021-68555.

6 Van Slyke, B., Mirkouei, A., and McKellar, M. (2021). Techno-economic and environmental assessment of dairy products: a case study in Southeast Idaho, USA. *Proceedings of ASME IDETC/CIE: 26th Design for Manufacturing and the Life Cycle Conference,* https://doi.org/10.1115/DETC2021-69285.

7 Thompson, M., Mohajeri, A., and Mirkouei, A. (2020). Environmental and economic impacts of nitrogen trifluoride at an Idaho semiconductor facility. *Proceedings of ASME IDETC/CIE: 25th DFMLC Conference.* https://doi.org/10.1115/DETC2020-22603.

8 Hersh, B., Mohajeri, A., Mirkouei, A. et al. (2020). Cyber-physical infrastructures for advancing pyrolysis conversion process: a case study of biochar production. *Proceedings of ASME IDETC/CIE: 25th DFMLC Conference.* https://doi.org/10.1115/DETC2020-22045.

9 Walters, J. and Mirkouei, A. (2020). Social life cycle assessment of computer-aided design tools. *Proceedings of ASME IDETC/CIE: 25th DFMLC Conference.* https://doi.org/10.1115/DETC2020-22576.

10 Hansen, S. and Mirkouei, A. (2019). Bio-oil upgrading via micro-emulsification and ultrasound treatment: examples for analysis and discussion. *Proceedings of the ASME IDETC/CIE: 24th DFMLC Conference.* https://doi.org/10.1115/DETC2019-97182.

11 Hersh, B. and Mirkouei, A. (2019). Life cycle assessment of pyrolysis-derived biochar from organic wastes and advanced feedstocks. *Proceedings of the ASME IDETC/CIE: 24th DFMLC Conference.* https://doi.org/10.1115/DETC2019-97896 **(One of three finalists for the Best Paper Award)**

12 Hansen, S., Mirkouei, A., and Xian, M. (2019). Cyber-physical control and optimization for biofuel 4.0. *Proceedings of the 2019 IISE Annual Conference H.E. Romeijn, A Schaefer, R. Thomas, eds.*

13 Hansen, S. and Mirkouei, A. (2019). Prototyping of a laboratory-scale cyclone separator for biofuel production from biomass feedstocks using a fused deposition modeling printer. *The Minerals, Metals & Materials Society 2019 Conference.* https://link.springer.com/chapter/10.1007/978-3-030-05861-6_26.

14 Hansen, S. and Mirkouei, A. (2018). Past infrastructures and future machine intelligence (MI) for biofuel production: a review and MI-based framework. *Proceedings of ASME IDETC/CIE: 23rd DFMLC Conference.* https://doi.org/10.1115/DETC2018-86150.

15 Mirkouei, A. and Kardel, K. (2017). Enhance sustainability benefits through scaling-up bioenergy production from terrestrial and algae feedstocks.

Proceedings of ASME IDETC/CIE: 22ⁿᵈ DFMLC Conference. https://doi.org/ 10.1115/DETC2017-67014.

16 Mirkouei, A., Silwal, B., and Ramiscal, L. (2017). Enhancing economic and environmental sustainability benefits across the design and manufacturing of medical devices: a case study of ankle foot orthosis. *Proceedings of ASME IDETC/CIE: 22ⁿᵈ DFMLC Conference.* https://doi.org/10.1115/DETC2017-68427.

17 Mirkouei, A. (2017). Enhance sustainability benefits through scaling-up bioenergy production from underutilized feedstock. *Proceedings of the 2017 IIE/ISERC.*

18 Mirkouei, A., Haapala, K.R., Murthy, G.S. et al. (2016). Reducing greenhouse gas emissions for sustainable bio-oil production using a mixed supply chain. *Proceedings of 2016 ASME IDETC/CIE: 21ˢᵗ DFMLC Conference.* https://doi.org/ 10.1115/DETC2016-59262 **(Received the DFMLC Scholar Award)**

19 Mirkouei, A., Bhinge, R., McCoy, C. et al. (2016). A pedagogical module framework to improve scaffolded active learning in manufacturing engineering education. *Procedia Manufacturing* 5: 1128–1142. https://doi.org/10.1016/j .promfg.2016.08.088.

20 Mirkouei, A., Haapala, K.R., Murthy, G.S. et al. (2016). Evolutionary optimization of bioenergy supply chain cost with uncertain forest biomass quality and availability. *Proceedings of the 2016 IIE/ISERC* (21–24 May), Anaheim, California, USA.

21 Mirkouei, A. and Haapala, K.R. (2015). A network model to optimize upstream and midstream biomass-to-bioenergy supply chain costs. *Proceedings of ASME 2015 International Manufacturing Science and Engineering Conference (MSEC).* https://doi.org/10.1115/MSEC2015-9355.

22 Mirkouei, A. and Haapala, K.R. (2014). Integration of machine-learning and mathematical programming methods into the biomass feedstock supplier selection process. *24ᵗʰ International Conference on Flexible Automation and Intelligent Manufacturing* (20–23 May) San Antonio, TX.

Appendix 3

Self-Check Questions and Answers

Chapter 1

1. What are the most used materials in PV panels? Crystalline silicon, cadmium telluride, copper indium gallium diselenide, perovskites, multijunction (iii–v), and organic.
2. What are the main techno-economic parameters in solar PV technologies? PV modules (37% of total cost), inverter (4–5% of total cost), electrical and structural BOS (21% of total cost), land, construction, engineering, and contingencies (20% of investment cost), operation and maintenance (2% of investment cost), interest rate, and insurance rate.
3. What are the benefits of flexible solar cells? (i) suitable for curved (uneven) surfaces, (ii) cheaper than silicon solar cells due to lower materials required, (iii) light weight, (iv) easy installation and handling, and (v) lower labor cost.
4. What is a crystalline silicon (C-Si) solar panel? And how does it work? The principles and science behind crystalline silicon (C-Si) solar panels are based on the physics of semiconductors and light properties. C-Si PV cells can be made of silicon wafers and treated with impurities to create a p–n junction (two different semiconductor types), allowing the cells to generate power from sunlight. The p-type semiconductor has positively charged holes, and the n-type semiconductor has negatively charged electrons. When the light photon is absorbed by the cells (silicon crystals), it can create electrical charge flows and can be used as a power source.
5. What are the top three used solar PV panels? And what are their benefits and drawbacks? Monocrystalline, polycrystalline, and thin film (Figure 1.5). The most efficient PV cells are monocrystalline, with 22% efficiency for up to 25 years, and they need less space compared to polycrystalline and thin film. The main drawback of monocrystalline cells is their high production cost, which makes them less cost-effective for cold regions. Polycrystalline cells are the most effective and sustainable method. The main drawback is

Net-Zero and Low Carbon Solutions for the Energy Sector: A Guide to Decarbonization Technologies,
First Edition. Amin Mirkouei.
© 2024 John Wiley & Sons, Inc. Published 2024 by John Wiley & Sons, Inc.

the low efficiency (up to 17%) due to high space requirements and the low silicon purity compared to monocrystalline cells. Thin-film cells have several benefits due to their lower weight compared to other cells, which makes them easy to move for remote applications. Their main drawback is low efficiency.

6. What are the required materials for C-Si PV? Metallurgical-grade silicon (MGS), glass, encapsulant film, backsheets, and aluminum frames.

7. What are the main benefits of multijunction PV cells? (i) spectrum matching with specific absorber layers and (ii) crystal structure and properties, such as compatible absorption spectra.

8. What are the main types of organic PV cells? (i) small-molecule cells with high absorption capacity in visible and near-infrared spectrum and (ii) polymer-based cells with high interface surface area.

9. What are the benefits and drawbacks of solar thermal towers for power and heat generation? The benefits are: (i) high thermal-to-power conversion efficiencies (30–40%), higher than many other types of renewable energy, (ii) energy storage that allows them to provide power even when the sun is not shining and makes them more reliable and dispatchable compared to some other renewable energy sources, (iii) scalability, which makes them suitable for a wide range of applications, (iv) low operating costs that can operate for many years with minimal maintenance, (v) no GHG emissions or other pollutants during operation, making them a clean energy source, and (vi) operating at high temperatures that can be used in industrial processes with high-temperature heat requirements. The main drawbacks are: (i) high land use to install the heliostats and tower that can be a challenge in areas where land is limited or expensive, (ii) high water consumption to generate steam and to cool the receiver that can be a significant constraint in areas with limited water resources, (iii) high capital costs than many other types of renewable energy, which can be a barrier to their deployment, (iv) technical complexity and a high level of expertise to design, build, and operate, which can add to the costs and risks associated with the technology, and (v) weather dependency, which can be impacted by weather patterns and seasonal changes and can affect their performance and reliability, and may require additional backup power sources.

10. What are the benefits and drawbacks of parabolic trough systems for power and heat generation? The benefits include: (i) achieving high thermal efficiencies due to the concentration of sunlight in a small receiver area, (ii) operating at high temperatures, suitable for industrial processes with high-temperature heat requirements, and (iii) being a sustainable energy source as they do not release GHG emissions or consume fossil fuels during operation, and their overall environmental impact is relatively low due to the manufacturing and installation of the components required for these systems. The main

drawbacks are: (i) the high cost of parabolic trough systems, (ii) receiver tube degradation, and (iii) oil-based heat transfer media that can limit output to moderate steam.

11. What are the main parameters that affect the performance of linear Fresnel reflector systems? The quality of the mirrors or reflectors, the design of the receiver, and the efficiency of the power conversion system.

12. What are the benefits and drawbacks of parabolic trough systems for power and heat generation? The benefits are high concentrations of sunlight, modular design, low water usage, and lower cost in comparison to other CHP systems, such as parabolic trough collectors. The main drawbacks are low efficiency, land use, and complex storage integration.

13. What are the main types of power generation from wind? Onshore, offshore, and airborne.

14. What are the benefits and drawbacks of power generation via onshore wind turbines? The main benefits are: (i) renewable and clean energy source, (ii) energy security and independence, (iii) minimal land use requirements, and (iv) cost-effective and low operational costs. The main drawbacks are: (i) visual and noise impacts, (ii) environmental impacts on wildlife and natural landscapes, (iii) wind speed variability, (iv) high capital costs, and (v) compatibility and integration challenges with existing energy systems.

15. What are the benefits and drawbacks of power generation via seabed-fixed offshore wind turbines? The main benefits are: (i) higher capacity and energy production than onshore wind power, (ii) no land use requirements, and (iii) low operational costs. The key drawbacks are: (i) high capital costs, (ii) complexity, (iii) potential environmental impacts on marine wildlife, (iv) potential impacts (e.g., visual and noise) for coastal communities, and (v) integration challenges with existing energy systems.

16. What are the benefits and drawbacks of power generation via floating offshore wind turbines? The main benefits include: (i) higher capacity than fixed offshore wind turbines, (ii) no land use requirements, and (iii) potential for larger wind farms. The main drawbacks include: (i) high capital cost, (ii) complex installation and maintenance requirements, (iii) potential impacts on marine wildlife and coastal communities, and (iv) integration challenges with other energy systems.

17. What are the benefits and drawbacks of power generation through hydropower? The benefits are: (i) renewable and clean energy source, (ii) high efficiency and reliability for continuous power generation, (iii) scalability, and (iv) other applications, such as irrigation or flood control. The main drawbacks are: (i) environmental impacts, including habitat destruction, water quality, aquatic ecosystems, and alteration of natural water flows; (ii) impacts on local communities and social structures; (iii)

water dependency that can be affected by droughts and other environmental factors; and (iv) high capital and maintenance costs, especially in remote or difficult-to-access areas.

18. What are the main geothermal technologies for power generation? Check Figure 1.23.

19. What are the benefits and drawbacks of geothermal power generation? Geothermal power generation has several benefits, including low emissions, reliable and dispatchable power, and local economic benefits. However, it has drawbacks, including high capital costs, limited availability, and potential negative impacts on local ecosystems and communities.

20. What are the benefits and drawbacks of flash steam geothermal power generation? The benefits are: (i) reliability and dispatchability, (ii) low operating costs, (iii) the ability to provide baseload power, and (iv) job creation and revenue generation for communities. The main drawbacks are: (i) the need for suitable geothermal reservoirs that can be limited to specific geographic areas, (ii) the construction and operation of the power plant and its associated infrastructure can also have some environmental impacts, and (iii) challenges associated with long-term sustainability and preventing depletion of geothermal reservoirs.

21. What are the key parameters affecting the performance of binary organic Rankine cycle geothermal power generation? Geothermal fluids' characteristics, temperature, and flow rate, as well as the efficiency of heat exchangers, turbines, and generators.

22. What are the benefits and drawbacks of binary organic Rankine cycle geothermal power generation? The benefits are reliability, stability, and sustainability, as well as operating at a wide range of geothermal fluid temperatures that make them suitable for a range of geothermal reservoirs. The systems are also compact and can be built in relatively small sizes. Binary ORCs have some drawbacks, such as geothermal fluid requirements and deployment barriers (e.g., high capital cost), as well as regular maintenance and replacement of equipment that can add to operational costs.

23. What are the benefits and drawbacks of enhanced geothermal systems for power generation? The benefits of EGS technology include the ability to: (i) work in areas with low to high-temperature geothermal resources, (ii) integrate with existing geothermal power generation systems or as a stand-alone technology, (iii) combine with other renewable energy sources to provide a reliable and consistent power supply, and (iv) operate continuously with minimal maintenance and downtime. EGS is secure, reliable, and sustainable to generate a significant amount of power, but its efficiency is currently lower than traditional geothermal power generation systems. The geothermal resource is replenished naturally over time, but the water used

in the process may contain impurities that need to be treated before disposal. Other drawbacks include that it requires special drilling techniques and is only economically feasible in areas with suitable geological conditions.

24. What are the main hydrogen generation methods? Steam-methane reforming, electrolysis, and gasification

25. What are the main steps for power generation through hydrogen-fired gas turbine systems? (i) air is drawn into the compressor, where it is compressed and heated, (ii) the compressed air is then mixed with hydrogen and burned in the combustor, (iii) the hot combustion gases expand through the turbine, where they drive the rotation of turbine blades, generating mechanical energy, and (iv) the exhaust gases then exit the turbine and are released into the atmosphere.

26. What is the operation principle of fuel cells? The operation principle of fuel cells is based on supplying hydrogen and oxygen that pass through the anode and cathode sides, respectively, where H^+ ions and electron release, and water are produced from mixing proton and oxygen (Figure 1.31). Fuel cells have different types and components that can facilitate the process, such as temperature or catalyst. The common catalyst is carbon materials coated with platinum, but the overall reaction is as follows: $2H_2 + O_2 \rightarrow 2H_2O + \text{electricity} + \text{heat}$.

27. What are the benefits and drawbacks of hydrogen-fired gas turbines for power generation? The benefits are high efficiency (around 60–65%), low emissions (no carbon, only water vapor), and low maintenance. The drawbacks are high cost (due to high hydrogen production cost), safety (hydrogen flammability), special handling (hydrogen storage and transportation).

28. What are the benefits and drawbacks of high-temperature fuel cells for power generation? The benefits are high efficiency (around 50–60%), low emissions, high scalability, and low maintenance. The drawbacks are high cost and short lifespan (5–10 years).

29. What are the benefits and drawbacks of hybrid fuel cell–gas turbine systems for power generation? The benefits are high efficiency (around 70–75%), low emissions, and high scalability. The drawbacks are high cost and system complexity.

30. What are the major marine technologies for power generation? Check Figure 1.32.

31. What are the main wave energy technologies for power generation? (i) floating- or submerged-point absorbers, (ii) shore- or floating-based oscillating water column, (iii) shore- or floating-based overtopping device, (iv) attenuator, and (v) oscillating wave surge converter (Figure 1.34).

32. What are the main steps during ocean thermal power generation? (i) Collecting warm surface water from the ocean using a pipe or other collection device,

(ii) Vaporizing a working fluid (e.g., ammonia or a fluorocarbon) in a heat exchanger to absorb heat energy from the warm water and vaporize into a gas, (iii) Running a turbine to produce mechanical energy that can be converted into power, (iv) Condensing the working fluid by cold water from the ocean depths to condense it back into a liquid that can be recycled back to the heat exchanger to repeat the cycle.

33. What are the main types of ocean thermal conversion technologies? Closed-, open-, or hybrid-cycle systems (Figure 1.35).

34. What are the basic steps for power generation from ocean salinity gradient? (i) Collecting seawater with different salinity levels from the ocean and stored in separate compartments, (ii) Using ion-exchange membranes that are placed between the compartments containing seawater with different salinity levels to selectively allow the passage of ions, creating a concentration gradient across the membrane, (iii) Generating electric potential difference as the ions move across the membrane, and (iv) Reversing the polarity to maintain the process and prevent the membranes from becoming saturated with ions.

35. What are the main ocean salinity gradient methods for power generation? (i) pressure retarded osmosis that can convert the osmotic pressure of saline solutions to hydraulic pressure to run the turbine and generate power, and (ii) reverse electrodialysis that uses cation and anion exchange membranes and different water salinities to transport ions through membranes, such as salt batteries and generate power (Figure 1.36).

36. What are the benefits and drawbacks of power generation from ocean salinity gradient? The main drawbacks are (i) the water requirement that can limit their usability and can impact local ecosystems and (ii) high capital costs, especially for ion-exchange membranes and semi-permeable membranes, and high operational costs for maintaining the concentration gradient across the membranes. The benefits include: (i) compatibility with existing power infrastructure, (ii) secure energy source as it is not subject to supply disruptions, and (iii) sustainable solution since it has minimal environmental impacts and does not release GHG emissions during the operation.

37. What are the main tidal power generation technologies? Tidal range and tidal stream.

38. What are solar hybrid energy systems? And what are their benefits? They use solar collectors with other renewable energy sources for power generation. The benefits include: (i) combining solar energy with other energy sources to ensure a stable and reliable power supply; (ii) using it in remote areas where there is no access to the grid or in areas where there is a high-power demand; and (iii) using as backup power sources for homes, businesses, and industries in case of power outages. The performance of these systems mainly depends on PV panel size, battery capacity, and inverter efficiency.

Chapter 2

1. What is the basic science of nuclear power generation? Nuclear-based energy comes from splitting uranium atoms in a process called fission to heat water into steam for powering steam turbines that can generate power (Figure 2.1).

2. What are the key parameters to control nuclear energy production? Temperature for reaction rate, water (the medium) for neutron absorber rate, and control rods for neutrons availability to the fissile materials.

3. What are the leading nuclear power generation technologies? Figure 2.2 presents the leading nuclear power generation technologies.

4. How does the light-water nuclear reactor work? Light-water nuclear reactors operate by using a controlled nuclear reaction to generate heat, which is then used to create steam to power turbines and generate power.

5. What are the benefits and challenges of light-water reactors for power generation? The benefits of light-water reactors include: (i) flexibility and compatibility with other power generation and transmission systems, (ii) high efficiency with minimal interruptions, and (iii) near-zero carbon energy source. The challenges are: (i) strict security protocols to prevent unauthorized access to nuclear materials, sabotage, and other security threats; (ii) vulnerability to weather or natural disasters; (iii) nuclear materials and wastes management, which are radioactive and require careful handling and disposal; and (iv) environmental impacts on aquatic ecosystems, as they require large amounts of water for cooling. These reactors are designed to be highly functional, generating large amounts of power with relatively low operating costs. However, they require significant maintenance and regulatory oversight to ensure safe and reliable operation. Light-water nuclear reactors are relatively easy to operate and maintain but require specialized knowledge and training. Additionally, handling nuclear materials and waste requires careful attention to safety and regulatory compliance.

6. How does sodium-cooled fast nuclear reactor work? Sodium-cooled fast reactors operate on the principle of nuclear fission, which generates heat that is used to produce steam and generate power. They are usually closed fuel cycle and fast neutron spectrum reactors, using liquid sodium as a coolant with an outlet temperature between 500 and 550 °C (Figure 2.4). The reactor can be arranged in a pool layout with different fuel options (e.g., mixed oxide and mixed metal alloy) and sizes, such as small (50–150 MWe) and large (600–1,500 MWe).

7. What are the main environmental impacts of sodium-cooled fast nuclear reactors for power generation? The environmental impacts are: (i) nuclear waste generation, which is radioactive and requires careful handling and disposal; (ii) significant impacts on aquatic ecosystems, as they require large amounts

of water for cooling; and (iii) high risk due to using radioactive materials that can cause long-term environmental damage.

8. What are the main advantages and disadvantages of sodium-cooled fast nuclear reactors for power generation? The advantages are low operation costs and low carbon emissions. The disadvantages are high risk and high capital costs. Sodium-cooled fast nuclear reactors are designed to be highly functional, generating large amounts of power with relatively low operating costs. However, they are more complex than light-water reactors and require more advanced technology and maintenance. These reactors are more challenging to operate and maintain than light-water reactors due to the highly corrosive nature of liquid sodium and the need for specialized training and expertise. Sodium-cooled fast reactors have high power density and low coolant volume fraction to reduce waste and increase yield.

9. How does a high-temperature nuclear reactor work? The science behind high-temperature nuclear reactors is based on nuclear fission, generating heat that can be used in combined energy systems for heat and power generation (Figure 2.5). These reactors are open-fuel cycle, helium-cooled, and thermal neutron spectrum reactors with outlet temperatures between 900 and 1,000 °C and capacity between 250 and 300 MWe.

10. What are the benefits of high-temperature nuclear reactors for power generation? The benefits of high-temperature nuclear reactors include: (i) highly functional for generating a high amount of power with low operating costs and minimal downtime; (ii) higher efficiency than other nuclear reactors due to high- and very high-temperature reactors that can produce both power and heat for several applications, such as water desalination or hydrogen production; (iii) flexibility and compatibility with other power and heat generation systems, and (iv) being a near-zero carbon energy source.

11. What are the challenges of high-temperature nuclear reactors for power generation? The challenges include: (i) special training requirements for both running the reactors and handling nuclear materials and waste, (ii) security concerns and environmental impacts due to radioactive materials, and (iii) high-temperature requirements and deployment barriers. The environmental impacts of high-temperature reactors are similar to those of other nuclear reactors, such as nuclear wastes, storage and disposal barriers, and large water requirements for cooling.

12. How does a small modular nuclear reactor work? Small modular reactors operate on the nuclear fission principle, using energy from controlled nuclear chain reactions to generate heat and power. These reactors have been under development for several decades, and they vary in size and capacity between 10 and 100 MW, using light water, gases, liquid metals, or salts as coolants.

13. What are the benefits of small modular nuclear reactors for power generation? (i) lower capital costs compared to other reactors; (ii) highly functional, modular, and scalable, allowing them to be easily deployed and scaled up as needed; (iii) portable and small physical footprints; (iv) high compatibility, flexibility, and efficiency compared to traditional reactors; and (v) zero or very low carbon emissions. The capital cost of developing a small modular power plant is lower than that of the traditional large-scale power plants due to the simplified design and size, about 1/10 to 1/4 of the size of a large-scale plant. Small reactors can be more efficient than large ones, but the operational cost of running the small reactors is higher than the larger reactors due to economies of scale.

14. What are the challenges of small modular nuclear reactors for power generation? (i) security and environmental concerns, (ii) high vulnerability due to their small size, and (iii) sustainability challenges, including nuclear materials and waste management. The main disadvantage compared to larger-scale power generation is the low overall power output.

15. What are the main environmental impacts of small modular nuclear reactors for power generation? The environmental impacts of small modular reactors are: (i) radioactive nuclear waste generation, which requires careful handling and disposal; (ii) impacts on ecosystems due to the water requirement for cooling; and (iii) potential risks due to the release of radioactive materials.

16. What are the differences between fission and fusion processes? Fission and fusion processes can produce massive energy through nuclear reactions than other sources from splitting larger atoms (fission) or joining smaller atoms (fusion). The fission process has been used mainly to heat water into steam for powering steam turbines that can generate power by splitting larger atoms into smaller ones. Currently, uranium and plutonium are mainly used for fission processes. However, the fusion process can generate energy when two lighter atoms slam together and form a heavier atom, such as two hydrogen atoms to form a helium atom, a process similar to what occurs in the sun.

17. What are the benefits of fusion reactors for power generation? The benefits include: (i) lower security risks, (ii) lower radiation risks compared to traditional nuclear reactors as they do not use fissile materials (e.g., uranium or plutonium), (iii) flexibility and compatibility with other power generation and transmission systems, (iv) sustainable method and energy source since they use hydrogen as fuel and generate no carbon emissions or radioactive waste, (v) cost-competitive energy source in the long term due to low fuel costs, using hydrogen as fuel that is abundant and easily accessible, and (vi) lower environmental impacts and wastes compared to traditional methods.

18. What are the drawbacks of fusion reactors for power generation? (i) environmental impacts due to high land use and energy requirements, (ii) high capital

and operational costs due to the complexity and other requirements (very high temperature and pressure), and (iii) difficulty in handling and maintaining due to the extreme conditions inside a fusion reactor. Fusion does not generate highly radioactive waste; however, it is difficult to run it for a longer time due to the high amount of pressure and temperature needed to form nuclei together.

19. What are the main steps in power generation from coal? Coal combustion, steam generation, cooling, and emissions control.

20. What are the most used, mature carbon-capturing technologies from coal, biomass, and natural gas? Check Figure 2.9.

21. What is the pre-combustion/physical absorption process for carbon capturing from coal? And how does it work? Pre-combustion/physical absorption process converts coal into a gas before it is burned, which allows for easier separation of carbon dioxide from the gas stream. The pre-combustion process involves gasification, water–gas shift reaction, and carbon capturing. This process consists of a gasifier to produce synthesis gas (syngas) and a water–gas shift reactor to produce CO_2 and hydrogen from the syngas (Figure 2.10). Gasification involves converting coal into a gas by reacting it with steam and oxygen in a high-pressure environment. This process produces a gas mixture known as syngas, which consists primarily of hydrogen and carbon monoxide. Water–gas shift reaction converts the carbon monoxide in the syngas into CO_2 and more hydrogen. This reaction requires the addition of water and a catalyst. After the water–gas shift reaction, the CO_2 capture unit separates CO_2 from the rest of the gas stream using solvent or adsorbent materials. The captured CO_2 can be compressed and transported for storage or utilization.

22. What are the benefits and drawbacks of the pre-combustion/physical absorption process for carbon capturing from coal? The pre-combustion process for CCUS from coal power plants can capture CO_2 before it is emitted into the atmosphere, making it another effective way to reduce carbon emissions. The pre-combustion process is highly efficient, with up to 90% of CO_2 captured from coal power plants. The main drawbacks of pre-combustion CCUS include: (i) complexity and high cost to implement due to special equipment and infrastructure and (ii) waste handling and management due to additional waste streams, e.g., coal ash and sulfur dioxide. This solution is a mature pathway for CO_2 capturing and has TRL 9, which needs further improvement to stay competitive. The pre-combustion process is much cheaper due to the less energy requirements compared to other CO_2 capturing methods. The main challenge is the overall efficiency that can be improved by different techniques, such as solvent regeneration (or avoiding solvent reduction) using ionic liquid.

23. What is the post-combustion/chemical absorption process for carbon capturing from coal? Post-combustion is another process used for carbon capturing from coal power plants, which involves capturing carbon dioxide from the flue gas produced by burning coal. Post-combustion technology uses a CO_2 absorber unit that contains sorbent (e.g., monoethanolamine solvent) and CO_2 stripper for separating the pure CO_2 from flue gas (Figure 2.11).

24. What are the main steps in the post-combustion/chemical absorption process for carbon capturing from coal? (i) capturing the flue gas (e.g., carbon, nitrogen, water vapor, and trace amounts of other gases) by burning coal; (ii) cooling the flue gas to reduce its temperature and increase the concentration of CO_2 in the gas stream; (iii) separating the carbon from the rest of the gas stream using solvent or adsorbent materials (e.g., amine); and (iv) compressing the CO_2 to reduce its volume and make it easier to transport, store, or utilize (e.g., enhanced oil recovery).

25. What are the main challenges of post-combustion CCUS technology for power generation from coal? (i) high energy requirement for capturing and separating CO_2 that can reduce the efficiency of the power plant; (ii) high cost of the equipment and infrastructure; and (iii) process complexity due to water condensation, selectivity reduction, and membrane temperature adjustment.

26. What are the most used post-combustion methods? Chemical absorption, physical adsorption, and membrane separation. Monoethanolamine chemisorption is the only commercialized method with high capital and operational costs. Membrane separation can enhance sustainability benefits and reduce operational costs.

27. What is the oxy-fuel combustion process for carbon capturing from coal? Oxy-fuel combustion is one of the main CCUS processes that involves burning coal in a mixture of pure oxygen and recycled flue gas, which creates a gas stream consisting primarily of carbon dioxide and water vapor. Carbon dioxide can then be captured and stored or utilized for other purposes. Oxy-fuel combustion technology uses a pure oxygen stream instead of air for combusting carbon-based fuel and separating CO_2.

28. What are the main steps of oxy-fuel combustion for carbon capturing from coal? The oxy-fuel combustion process for carbon capturing involves several steps (Figure 2.12). The first step is air separation into its component gases, primarily nitrogen and oxygen, using an air separation unit that relies on cryogenic distillation or other separation techniques. The second step is coal combustion in a mixture of oxygen and recycled flue gas. The recycled flue gas is used to provide the necessary volume of gas for combustion and to moderate the temperature of the combustion process. The combustion process generates a gas stream, consisting primarily of CO_2 and water vapor. The third step

is CO_2 capturing and separation from the gas stream, using a solvent or adsorbent material similar to post-combustion processes.

29. What are the main components of the oxy-fuel combustion process for carbon capturing from coal? Oxygen separator, boiler, particle removal, condenser units, and compressor that can combust carbon-based fuel, using oxygen and send the flue gas to particle removal and condensers to remove water and sulfur and separate CO_2.

30. What are the benefits and drawbacks of the oxy-fuel combustion process for carbon capturing from coal? One of the advantages of oxy-fuel combustion is that CO_2 is already concentrated in the gas stream, making it easier to capture and separate from other gases. Then CO_2 is compressed to high pressure, transported to a storage site, or utilized for other purposes. The oxy-fuel combustion process is straightforward, but the main challenges are the energy required to separate air into its component gases and to compress CO_2. Also, the costs of required equipment and infrastructure are relatively high. The energy-intensive, high-cost process (e.g., cryogenic distillation) for pure oxygen production is the limiting factor, but overall oxy-fuel technology is the energy-efficient pathway for CO_2 separation and capturing. Despite these challenges, oxy-fuel combustion is an effective method for reducing CO_2 emissions from coal power plants. It can achieve high levels of CO_2 capture (up to 90% or more), and the captured CO_2 is already compressed, reducing the energy required for transport and storage.

31. What is the chemical looping combustion process? And how does it work? Chemical looping combustion is another CCUS approach for reducing carbon emissions from coal power generation. This solution can capture CO_2 after burning coal without energy-intensive separation processes. The basic science behind chemical looping combustion for CCUS involves using metal oxide (MeO) particles (e.g., iron or nickel) as the catalyst. In this process, MeO is circulated between the fluidized reactors as the oxygen carrier to produce CO_2 and water in the reducer (Figure 2.13). Later, methyl (Me) can be separated in the cyclone and reused in other cycles for reducing the process cost.

32. What are the main components of the chemical looping combustion process for carbon capturing from coal? This process uses two reactors (i.e., air and fuel) for circulating MeO particles. In the air reactor, MeO particles are exposed to air to oxidize and release heat. In the fuel reactor, hot MeO particles react with the coal to generate heat and CO_2, and capture CO_2 by MeO particles. Then loaded MeO particles with CO_2 are transported back to the air reactor to start the cycle again. The captured CO_2 can be removed from MeO particles by heating them to a high temperature, which releases CO_2 and regenerates MeO particles.

33. What are the benefits and challenges of the chemical looping combustion process for carbon capturing from coal? The benefits of chemical looping combustion include: (i) high CO_2 capturing level without energy-intensive separation processes and (ii) lower oxygen concentration than traditional combustion processes, which reduces nitrogen amount in the flue gas and simplifies CO_2 capture. The challenges include: (i) the high cost and complexity of the reactor system, which requires two reactors and a circulating fluidized bed and (ii) the limited availability of suitable MeO particles, which can be expensive and difficult to produce. Earlier techno-economic studies reported that chemical looping combustion is more cost-effective than oxy-fuel combustion.

34. What is utilization in CCUS strategies? Utilization is one of the main components of CCUS technologies, which involves the conversion of captured CO_2 into valuable products, rather than simply storing it underground.

35. What are the most used utilization applications? Currently, the largest end use for carbon capturing is enhanced oil recovery (EOR). EOR is a process that involves injecting CO_2 into oil reservoirs to increase oil production. The injected CO_2 dissolves into the oil through a process called solubility trapping, reducing its viscosity, and making it easier to extract. The solubility of CO_2 in oil is influenced by pressure, temperature, and the composition of the oil. It can be an attractive utilization option for CCUS from coal power plants because it can generate revenue from the sale of the additional oil produced. Other applications include chemical or fuel production, beverage carbonation, mineral carbonization, metallurgy extraction, and machinery. Production of chemicals and fuels is another potential utilization option for CCUS from coal power plants. Captured CO_2 can be converted into chemicals (e.g., methanol) or fuels (e.g., synthetic diesel) through a carbon capture and utilization process that can be an attractive option to generate revenue from byproducts. The captured CO_2 is converted into chemicals or fuels through various processes, such as catalytic conversion, electrochemical conversion, or biological conversion. The specific process used will depend on the desired product and the characteristics of the feedstock. Mineralization is a process that involves the reaction of captured CO_2 with naturally occurring minerals (e.g., magnesium and calcium silicates) to form stable carbonates. The captured CO_2 reacts with minerals in the presence of water to form carbonates. The reaction kinetics and product formation are influenced by mineral composition, reaction temperature and pressure, and the presence of catalysts. Mineralization can be an attractive utilization option for CCUS from coal power plants because it permanently removes CO_2 from the atmosphere and can potentially generate revenue from the sale of the resulting carbonates.

36. What are the advantages of power generation from natural gas? (i) a relatively clean-burning fossil fuel, emitting less pollutants (e.g., sulfur dioxide, nitrogen

oxides, and particulate matter) compared to coal or oil; and (ii) a higher thermal efficiency compared to other fossil fuels for power generation, meaning it can generate more power using the same amount of fuel.

37. What is the main disadvantage of power generation from natural gas? And what are the solutions to address it? The major disadvantage is that the combustion of natural gas still produces carbon dioxide and other GHGs that contributes to climate change. To mitigate the environmental impacts of natural gas power generation CCUS technologies can be used similar to coal power plants to capture and sequester carbon emissions (Figure 2.15). There are several approaches that have been investigated, and the main ones are post-combustion/chemical absorption and supercritical CO_2 cycle.

38. What are the main carbon capturing, transportation, and storage parameters? The main carbon capturing parameters are the type of capture technology used, the temperature and pressure of the flue gas, and the type of solvent or sorbent used. The main transportation parameters are the distance between the capture site and the storage site, the type of transportation method used, and the safety and security of the transport process. The main storage parameters are geological formation types used for storage, CO_2 injection rate, and the long-term stability of stored CO_2.

39. What are the main steps of the post-combustion/chemical absorption process for carbon capturing from the natural gas power plant? This process involves capturing CO_2 emissions from the flue gas using a solvent, such as an aqueous amine solution (Figure 2.16). The main steps include: (i) the flue gas from the natural gas power plant goes through an absorber tower for spraying the solvent into the gas stream; and (ii) the solvent reacts with CO_2 in the gas, forming a chemical bond and removing CO_2 from the gas stream. For example, the amine solvent acts as a weak base, reacting with acidic CO_2 to form a chemical bond. The chemical reaction is reversible, so the solvent can be regenerated by heating it and releasing CO_2, which is then stored or used for other purposes.

40. What are the main parameters in the post-combustion/chemical absorption process for carbon capturing at the natural gas power plant? The key physical interaction parameter in the post-combustion process is the CO_2 mass transfer rate from the flue gas into the solvent. The mass transfer rate depends on the CO_2 concentration in the flue gas, gas temperature and pressure, and solvent properties, such as viscosity and surface tension. The efficiency of the post-combustion/chemical absorption process depends on gas concentration and flow rate, solvent type and flow rate, and absorber tower operating conditions (e.g., temperature and pressure).

41. What is the supercritical CO_2 cycle? And how does it work? The supercritical CO_2 cycle is a novel technology that can be used for CCUS from natural gas

power plants. The basic science of this technology involves the thermodynamics of a fluid in a supercritical state and the operation of a closed-loop cycle. Supercritical CO_2-based power cycles operate similarly to other steam-based power cycles; however, they use CO_2 as the working fluid. During power generation from natural gas, the exhaust gas from the turbine is directed through a heat exchanger, where it transfers its heat to a supercritical CO_2 fluid, which is maintained at high pressure and temperature (above its critical point). Figure 2.17 shows both directly and indirectly heated supercritical CO_2-based power cycles.

42. What are the main carbon-capturing methods applied to natural gas power plants? And what are their benefits and drawbacks? Post-combustion/chemical absorption, supercritical CO_2 cycle, and oxy-fuel combustion. Post-combustion/chemical absorption process is the most common CCUS method due to its simplicity, low energy consumption, and compatibility with existing power plants, but the drawbacks are high capital costs, solvent requirements, and wastewater generation. The benefits of the supercritical CO_2 cycle process include its high efficiency, low water consumption, and small footprint. The main drawback is the need for advanced materials for a supercritical (high pressures and temperatures) process. The benefits of the oxy-fuel combustion process include its high purity of captured CO_2, compatibility with existing power plants, and relatively low capital cost. The drawbacks are oxygen requirement and corrosive gas stream.

43. What are the main steps for power generation from biomass feedstocks? And how does each step work? Biomass preparation, conversion process, steam generation, and cooling. In particular, biomass feedstocks are first collected, sorted, and processed to remove contaminants, such as rocks, soil, and noncombustible materials. The feedstocks are then shredded, chipped, or ground into a uniform size for efficient combustion. The pretreated biomass feedstocks, after dewatering and size reduction, are burned in a boiler to produce high-pressure steam. The thermochemical (combustion) conversion process can be either direct combustion, pyrolysis, or gasification, depending on the type of biomass and the desired energy output. The steam generated from the thermochemical process is directed to a steam turbine, which is connected to a generator. The steam turbine converts the thermal energy of steam into mechanical energy, which drives the generator to produce power. The steam that passes through the steam turbine is then cooled using a condenser (cooling tower) that can convert the steam back into water. The water is then returned to the boiler to be reheated and reused in the steam generation process.

44. What are the main steps of the post-combustion/chemical absorption process for carbon capturing from biomass power generation? (i) flue gas treatment

for removing particulate matter, sulfur dioxide, and other impurities from biomass combustion and reducing viscosity and corrosion; (ii) CO_2 absorption, using an absorption tower and solvents (e.g., monoethanolamine or ammonia) that can react with CO_2 to form a chemical compound and separate it from the flue gas; (iii) CO_2 separation from the CO_2-rich solvent, using the stripping column, while the solvent is recycled back to the absorption tower; and (iv) CO_2 compression to a supercritical state and transport it to storage sites (e.g., depleted oil and gas reservoirs or saline aquifers).

45. What are the main steps of the pre-combustion/physical absorption process for carbon capturing from biomass power generation?

Gasification: The biomass feedstock is converted into syngas through gasification, which involves heating the biomass in the presence of a gasifying agent (e.g., oxygen, steam, or air). The syngas consists of carbon monoxide, hydrogen, and other impurities.

Shift conversion: The syngas is then subjected to a shift conversion process, where CO is reacted with steam to produce more H_2 and CO_2. This step increases the concentration of CO_2 in the syngas and prepares it for CO_2 capture.

CO_2 capture: CO_2-rich syngas is directed to a physical absorption column, where it is treated with a solvent (e.g., Selexol or Rectisol) to capture CO_2. The solvent reacts with CO_2 to form a chemical compound that can be separated from syngas.

Separation: CO_2-rich solvent is directed to a stripping column, where heat is applied to separate CO_2 from the solvent. Then CO_2 is compressed and transported for storage or utilization, while the solvent is recycled back to the absorption column.

Power generation: The clean syngas is then used to fuel a gas turbine or internal combustion engine to generate power. The generated power can be used on-site or fed into the grid.

46. What are the benefits and drawbacks of the pre-combustion/physical absorption process for carbon capturing from biomass power generation? The benefits include: (i) high CO_2 capture rate (up to 95%), (ii) low energy requirements, (iii) higher efficiency due to the use of byproducts (e.g., syngas) for other applications (e.g., heating or transportation fuel), and (iv) the potential for co-production of hydrogen. The drawbacks include: (i) high water and solvent needs, (ii) high potential for solvent degradation and corrosion, and (iii) complex equipment and a higher capital investment, which may limit its scalability.

47. What are the main technologies for power generation from ammonia? Ammonia turbines, cracking into hydrogen for gas turbines, and co-firing in coal power plants are mature pathways using ammonia as a fuel source for power

generation, but they differ in their basic operating principles and efficiency (Table 2.4).

48. What are the benefits of power generation from ammonia? (i) ammonia can be produced from renewable energy sources, such as wind or solar power, making it a renewable energy carrier; (ii) ammonia produces nitrogen and water vapor as byproducts, which are environmentally friendly and do not contribute to GHG emissions; (iii) ammonia has a high energy density, which means it can store more energy per unit volume than other fuels, such as hydrogen or methane; (iv) the existing infrastructure can be used for ammonia production and transport, which can be leveraged for power generation; and (v) ammonia can be used for power generation in a variety of ways, including combustion, fuel cells, and synthesis gas production.

49. What are the drawbacks of power generation from ammonia? (i) ammonia is toxic and can be dangerous if not handled properly, which can be a challenge for large-scale production and transportation; (ii) ammonia is highly volatile and requires specialized storage and transportation infrastructure to prevent leaks and accidents; (iii) ammonia has high ignition energy, which can make it challenging to use as a fuel in specific combustion engines and gas turbines; (iv) the production and transportation of ammonia can be costly, and the cost of ammonia-based power generation technologies may be higher than conventional fossil fuel-based power generation; and (v) many of the ammonia-based power generation technologies, such as ammonia fuel cells, are still in the development stage, which may delay their commercialization and widespread adoption.

50. What are the key carbon removal technologies? Check Figure 2.21.

51. What are the examples of CCUS hybrid energy systems for power generation? Integrated gasification combined cycle (IGCC) with carbon capture, supercritical pulverized coal with carbon capture (SCPC), and natural gas combined cycle (NGCC) power plant with carbon capture.

52. What are the benefits and challenges of CCUS hybrid energy systems for power generation? The benefits are: (i) reduced carbon emissions by capturing and storing CO_2 emissions underground, (ii) increased efficiency of power generation by reducing energy loss associated with CO_2 capture and storage, and (iii) economic benefits by creating new industries and job opportunities. The challenges include: (i) high cost compared to traditional fossil fuel-based power generation technologies, which is mainly due to the high cost of capturing and storing CO_2 emissions, (ii) limited storage capacity since underground storage capacity for large volumes of CO_2 is limited, (iii) energy loss and reduced efficiency of power generation, and (iv) technological barriers to deploy on a large scale.

Chapter 3

1. What are the main types of power storage technologies? Existing commercialized power storage technologies can be classified into two main categories: batteries and mechanical energy storage. Battery technologies have unique features, e.g., size, shape, and portability. There are several different technologies for power storage, each with its own benefits and drawbacks (Figure 3.1). Also, chemical storage technologies (e.g., hydrogen and ammonia) have the potential for energy storage and convert it to power in the future.

2. What are the critical parameters in power storage technologies? Several parameters play a crucial role in power storage, such as efficiency, lifespan (years), life cycle, energy density (Wh/L), capacity (MW or GW), safety, and cost. Efficiency measures how much power is lost or can be stored and recovered from the power storage system compared to the power used to charge and discharge the system. The life cycle measures how often the storage system can be adequately used before losing its ability to store power. Energy density estimates the amount of power that can be saved per volume (Wh/L). Higher efficiency means less power loss, and higher energy density means more energy storage in a smaller device. For storage systems, safety is critical, especially in high-capacity storage systems. For example, storage systems must be designed to prevent overheating and explosions, especially due to external temperature changes. Cost is another critical parameter in power storage, which determines commercialization viability, makes them more accessible and affordable, and increases widespread adoption.

3. What are the benefits and drawbacks of batteries for storing power? Batteries are electrochemical power storage systems with numerous benefits, such as being portable and easy to use, as well as high energy efficiency and density. The main drawbacks of batteries are short lifetime and high capital costs, along with high environmental impacts (e.g., hazardous wastes) due to their raw materials, including metals or non-metals. Earlier studies show that the grid-scale (MW) battery has significant environmental impacts.

4. What are the primary commercialized batteries? Currently, lithium-ion batteries are the most widely used energy storage system, covering over 90% of the market in 2020 (Figure 3.3). Lithium-ion and vanadium redox flow batteries have the highest TRL. Other types are sodium–sulfur (Na–S), nickel–cadmium (Ni–Cd), and lead–acid (Pb–A). Battery lifetime is 5–20 years, and their efficiency is around 65–90%.

5. What are the key parameters that affect the performance of batteries? (i) electrode materials (e.g., lithium cobalt oxide, nickel–cadmium, and lead–acid) that greatly affect the battery performance, including its capacity, voltage, and lifespan; (ii) electrolyte (e.g., lithium salt solutions, sulfuric acid, and

potassium hydroxide) that is the medium through which ions move between the electrodes, and it also affects the battery performance; (iii) separator, which is a permeable membrane that separates the two electrodes, allowing ions to pass through while preventing direct contact between the electrodes, and affects the battery capacity, internal resistance, and safety; (iv) temperature that affects the rate of chemical reactions inside the battery and can greatly impact its performance and increase the battery capacity but can also shorten its lifespan; (v) charging and discharging that affects their capacity, voltage, and lifespan; and (vi) battery management system that can monitor and control the battery charge and discharge rates, temperature, and other factors to improve its performance and lifespan.

6. How does a lithium-ion battery work? The basic science behind lithium-ion batteries involves electrochemical reactions, using lithium ions to generate power. Particularly, lithium ions move from the cathode to the anode during charging, where they are stored and the ions return to the cathode during discharging, which creates a flow of electrons and generates power.

7. What are the main components of lithium-ion batteries? The main components are two electrodes (cathode and anode), an electrolyte, and a separator. The primary materials used are lithium cobalt oxide (27%), steel (20%), graphite (16%), and polymer (14%). Most existing lithium-ion batteries are made of the positive cathode electrode from lithium cobalt oxide, the negative anode electrode from graphite, and the electrolyte from a lithium salt dissolved in an organic solvent, as well as the separator that is made of a thin, porous material to physically separate the cathode and anode while allowing lithium ions to move between the electrodes during charging and discharging.

8. What are the benefits of lithium-ion batteries? Lithium-ion batteries have high efficiency and energy density of up to 97% and 500 Wh/L, respectively. They can also last over 1,000 charge cycles. The main benefits of these batteries are: (i) high energy density to store more power in small size and weight, which makes them popular for use in portable devices and electric vehicles; (ii) long lifetime, depending on several factors, such as material quality, applications, and charging and discharging conditions; and (iii) rapid charging and stability in various temperatures.

9. What are the drawbacks of lithium-ion batteries? The main drawbacks include: (i) capacity degradation over time due to natural chemical reactions and extreme temperature changes, (ii) environmental impacts due to toxic chemicals (e.g., lithium and cobalt) and high GHG emissions from the production process, (iii) expensive and complex handling and recycling requirements, and (iv) safety issues, such as internal short circuits or fire.

10. What are the main environmental impacts associated with lithium-ion batteries? The main impacts are from: (i) resource extraction, (ii) energy-intensive

production process, (iii) limited recycling or disposal infrastructure, and (iv) safety concerns.

11. How does a redox flow battery work? And what are the main components of these batteries? The basic science behind redox flow batteries involves electrochemical reactions, using liquid electrolytes that flow through the battery during charging and discharging. These batteries are made of two different electrolytes, separated by a membrane, to store and release energy. The electrolytes are from metal ions in various oxidation states (e.g., vanadium, iron, or zinc). The electrolytes are stored in separate tanks and pumped through the battery during operation. The oxidation state of the metal ions determines the battery charge state. The membrane allows the flow of ions, but it prevents the mixing of the two electrolytes.

12. What are the key parameters of redox flow batteries? Electrolytes, membrane, efficiency, energy density, and durability. The electrolytes' characteristics and membrane types can improve efficiency and performance, as well as reduce degradation over time.

13. What are the benefits of redox flow batteries? Large-scale energy storage applications, stability, low degradation, and long lifetime compared to other batteries.

14. What are the drawbacks of redox flow batteries? (i) low energy and power density (25–35 Wh/L), (ii) high capital cost compared to lithium-ion batteries, (iii) environmental impacts due to mining and processing high metal requirements (e.g., vanadium, iron, and zinc), (iv) high energy requirements, and (v) large-scale infrastructure requirements, such as tanks for storing the electrolytes, pumps for circulating the electrolytes, and power electronics for controlling the flow of energy.

15. What are the main environmental impacts associated with redox flow batteries? The main impacts are from the following: (i) mining operations that contribute to soil erosion, water and air pollution, and habitat and land destruction; (ii) energy-intensive processes that contribute to GHG emissions; and (iii) toxic materials (e.g., vanadium and sulfuric acid) that can pose environmental risks.

16. How does mechanical energy storage work? Mechanical energy storage technologies use kinetic (energy of motion), weight, or gravitational forces for saving energy.

17. What are the most used mechanical energy storage systems? The commercialized and most used mechanical energy storage systems are pumped hydropower, flywheel, liquid air, and compressed air energy storage.

18. What are the differences between batteries and mechanical energy storage systems? The main difference between batteries and mechanical energy

storage technologies is their operational and scientific concepts (e.g., materials science, electrochemistry, and thermodynamics). Particularly batteries work by utilizing electrochemical reactions, and mechanical energy storage technologies work by using kinetic energy for storing and releasing power. Compared to batteries, mechanical energy storage technologies do not require toxic chemicals (e.g., lithium and cobalt) that negatively impact the environment. They can be built from recycled steel and have a longer life with fewer replacements.

19. How does pumped hydropower store energy? Pumped hydropower energy storage technology can store energy and generate power from two water reservoirs at different elevations. The science behind pumped energy storage involves storing energy in the form of potential energy, using water and gravity like a giant battery. Particularly, the energy is used to pump the water to a higher elevation (charge) and run the turbines when water moves down (discharge) to generate power.

20. What are the main types of pumped hydropower energy storage systems? And what are the main differences? Open-loop and closed-loop (Figure 3.5). The main difference is the connection to the natural body of water (e.g., rivers or lakes). The closed-loop reservoirs are not connected to the outside or natural water, and most of the new plants in the United States. are closed-loop. The main challenge is to find locations (topographies) with significations (low and high) in close proximity. There are other types, such as pumping water into tanks full of gas or air to compress and pressurize the gas and open it to push the water and run the turbine when the grid needs power.

21. What are the key parameters that affect pumped hydropower energy storage systems? Efficiency, capacity, elevation, location, infrastructure costs, and environmental impacts.

22. What are the advantages and disadvantages of pumped hydropower energy storage systems? The main advantages are high capacity, low cost per kWh, long lifetime, and reliability. The disadvantages are geographical limitations, high capital costs, and environmental concerns.

23. How does flywheel energy storage work? And what are the main parameters of this technology? Flywheel is one of the earliest energy storage technologies, using a rotating mechanical device with widespread applications (e.g., aerospace and telecommunications) to store energy (MW) and discharge in a few minutes. The science behind flywheel energy storage systems involves storing kinetic energy in rotational motion. The key scientific parameters are rotational motion, inertia, and friction or losses from air resistance or bearing friction.

24. What are the main factors affecting the performance of flywheel energy storage technology? Rotor mass, rotor speed, bearing friction, and materials.

Higher rotor mass or speed can increase the amount of stored energy. The bearing friction from supporting the heavy rotor (wheel) can reduce the efficiency and durability of the system by losing power.

25. What are the main components of flywheel energy storage technology? (i) a vacuum chamber for reducing the air resistance and maximizing the efficiency of the flywheel, (ii) a cylinder with a large mass that spins at a substantial speed of several thousand RPM, (iii) a motor for storing energy by spinning the flywheel and recovering the stored energy by spinning back and acting as a generator, (iv) magnetic bearings for supporting the flywheel and reducing the mechanical wear and frictional losses. The rotating cylinder can be made out of steel or carbon composite. The composite is lighter and stronger, and can achieve higher rotational speeds.

26. What are the benefits and drawbacks of flywheel energy storage technology? The benefits are low maintenance, long lifetime, high power density, fast response times, and low environmental impacts during operation. The main drawbacks are high capital costs, limited energy storage duration (seconds to minutes), and the need for sophisticated control systems because mechanical failure can be catastrophic due to massive forces or overheating. Flywheel energy storage technology is more appropriate for short-duration, high-power applications.

27. What are the main environmental impacts of flywheel energy storage technology? Material extraction and processing, such as steel, copper, and rare earth metals. Other impacts include high noise and vibration during the operation.

28. How does liquid air energy storage work? The basic science behind liquid air energy storage systems involves compression, cooling, and energy release. Liquid energy storage liquefies the air in very low temperatures at around $-200\,°C$ (cryogenic), stores it in a tank, then brings it back to the gas state with ambient air or waste heat from other industrial processes, and uses the gas to turn a turbine and generate power.

29. What are the main parameters of liquid air energy storage technology? The key parameters that affect the performance of these storage systems include: process efficiency, thermal insulation, turbine design, and size. Improving these parameters reduces energy loss and increases performance.

30. What are the benefits and drawbacks of liquid air energy storage technology? The benefits are a long lifetime, large-scale storage capacity (GWh), power outputs (100 MW), low environmental impacts, and fewer safety challenges. The main drawbacks are the high cost of manufacturing and installation, and limited energy storage duration (hours to days).

31. How does compressed air energy storage work? The basic science of compressed air energy storage is similar to liquid air energy storage, involving compression, storage, and energy release. It uses ambient air or another

gas to compress and store energy under high pressure (around 1,000 psi) in underground containers (Figure 3.7). It is very similar to pumped storage in terms of plant size and applications, and has the potential for both small- and large-scale storage needs. The main methods for compressed air energy storage are diabatic or adiabatic. In adiabatic, the heat from air compression is stored, recovered, and reused to reheat the compressed air and increase efficiency by up to 70%.

32. What are the key parameters that affect the efficiency and performance of compressed air energy storage? Capacity, turbine design, and air quality.
33. What are the benefits and issues of compressed air energy storage systems? High capacity, low cost per kWh, low storage losses, and long storage periods (days). The issues are high capital cost, low energy density, and site selection, which requires large and suitable geology to store compressed air underground.
34. What are the main environmental impacts of compressed air energy storage systems? High land use, air pollution, noise, and vibration during the operation.

Chapter 4

1. What is the basic science behind heat generation? The basic science behind heat generation involves heat transfer mechanisms, including conduction, convection, or radiation.
2. What are the key parameters and the primary demand for heat generation? The key parameters are the heat sources (e.g., fossil fuels, geothermal, or solar radiation) and the input–output energy ratio. The heat demand in commercial and residential buildings accounts for over 60% of the total energy needed in cold regions and over 30% in warm regions.
3. What are the major applications and needs for heat generation? The major industrial applications of heat generation are metal processing, distract heating, crude oil heating, food processing, pharmaceutical, and chemical industries. Heat generation technologies highly depend on the heat source (i.e., renewable or non-renewable).
4. What are the main non-renewable (fossil fuel) based heat generation technologies? And what are their advantages and disadvantages? Natural gas-fired or coal-fired boilers and furnaces. They have been used widely due to their high energy density, however, their cons include environmental impacts, such as air pollution from fossil fuel combustion, water, and soil pollution from waste discharge, along with ecosystem disruption due to resource extraction.

5. What are the main renewable resources for heat generation technologies? And what are their pros and cons? The primary renewable sources for heat generation are solar thermal, geothermal, and biomass. They can minimize the environmental impacts using renewable sources, but they may require higher capital costs due to specific design and infrastructure.

6. What is solar thermal district heating technology? And how does it work? Solar thermal district heating technology uses solar-based energy to generate and distribute heat to a network of buildings. Solar-based energy can be generated everywhere, especially in deserts, remote, or low-income regions. The basic science behind solar thermal district heating technology involves capturing solar radiation with solar collectors to heat up the transfer fluid and then circulating the heat through the system.

7. What are the main components of solar thermal district heating technology? Solar collectors, heat storage tanks, heat exchangers, and distribution pipes.

8. What are the main types of solar thermal district heating technology? This technology can be classified into two types: (i) centralized systems with a central solar collector and storage and (ii) decentralized systems with distributed solar collectors (e.g., roof space) and without storage (Figure 4.1). Centralized systems with ground-mounted solar collectors and seasonal storage have been fully commercialized, but decentralized systems have not been commercialized yet.

9. What is the science behind heat generation through heat pumps? The basic science behind heat pumps involves generating or transferring heat from a low-temperature source to a high-temperature sink. They use the thermodynamic cycle principle to warm or cool the house and provide a comfortable temperature.

10. What are the main types and applications of heat pumps? The main types are air-source or ground (geothermal)-source heat pumps that can provide up to 160 °C hot air and steam for various process needs (Figure 4.2). Heat pump technologies can be used for various applications, such as heating and cooling for buildings or industrial purposes.

11. What are the main types of large-scale heat pumps? Large-scale heat pumps can be classified based on their temperature level and refrigerant, such as standard (up to 80 °C), high (up to 100 °C), or very high (up to 160 °C) temperature heat pumps with synthetic–organic refrigerants (e.g., hydrochlorofluorocarbon, hydrofluorocarbon, or hydrofluoroolefin) or with R717 refrigerant.

12. What are the main components of large-scale heat pumps? Compressor, evaporator, condenser, and expansion units.

13. What are the key parameters of large-scale heat pumps? Heat pump's efficiency (the ratio of heat output to energy input), capacity, and temperature range.

14. What are the benefits and drawbacks of large-scale heat pumps? Their main benefits are a long lifetime, low operational costs, low maintenance, low emissions, and high safety compared to other conventional heat generation technologies, such as coal-fired boilers. However, their drawbacks are high capital costs for the main components and potential refrigerant leaks.

15. How does a geothermal heat pump work? The basic science behind geothermal heat pumps involves using the constant Earth's temperature throughout the year because the temperature at about 10 m below the surface remains relatively stable between 10 and 15 °C. Figure 4.3 presents community-scale heating and cooling system, using geothermal boreholes up to 150 m deep. Geothermal heat pumps can take advantage of the Earth as a heat source and sink (thermal storage) to efficiently exchange temperatures from underground for heating and cooling in winter and summer.

16. What are the main components of the geothermal heat pump for heat generation? Heat pump, underground loop (series of pipes), and fluid that can circulate in the loop by absorbing heat from the ground for heating or absorbing heat from the building for cooling. The loop can be developed either horizontally or vertically.

17. What are the benefits and drawbacks of geothermal heat pumps? Similar to large-scale heat pumps, the main benefits are high efficiency, low operational costs, low energy use (up to 60%), and low emissions. The drawbacks are high capital costs and high land use.

Chapter 5

1. What are the main types of energy storage systems? (i) power storage, such as batteries for small-scale or mechanical storage methods for large-scale, and (ii) heat or thermal energy storage that can store energy from the sun, geothermal, or heat waste.

2. What are mature heat storage technologies that have been used on a large scale? Latent, sensible, and thermochemical heat storage (Figure 5.1).

3. What are the main parameters to compare different heat storage technologies? Storage energy density, efficiency, heat capacity, storage period, cost, safety, and temperature range.

4. How does sensible heat storage work? It is based on increasing the temperature of an element or material (charging) and recovering the energy by dropping its temperature (discharging). Water is the widely used material for low temperatures (below 100 °C) due to its availability and nontoxic characteristics. For high temperatures (above 100 °C), solid materials can be used for heat storage, such as ceramics and concrete.

5. What are the benefits and drawbacks of using molten salt for high-temperature heat storage systems? Molten salt has been widely used for high-temperature heat storage systems due to its low costs, availability, and chemical characteristics and attributes, such as high heat capacity, temperature stability, boiling point, and nonflammability. Molten salt can act as heat transfer and storage material, but the drawbacks are high melting point, viscosity, and low thermal conductivity.

6. What are the common sensible heat storage media? Aluminum oxide, carbonate salts, hydroxide salts (e.g., NaOH), graphite, sodium chloride (NaCI), nitrate salts, and sodium liquid metal. Table 5.1 compares the thermophysical properties of sensible heat storage media.

7. What are the attributes of sensible heat storage technology? Small energy density storage (15–$50\,kWh/m^3$), medium/high efficiency (50–90%), and small capacity (10–$50\,kWh/t$).

8. What are the main parameters of sensible heat storage? Material mass, specific heat capacity, and temperature range.

9. What is specific heat capacity? The amount of heat required to increase the temperature of a material (e.g., solid or liquid) to a certain amount.

10. What are the benefits and drawbacks of sensible heat storage technology? The benefits are high efficiency, low cost, no emissions during the operation, and a wide range of applications. However, sensible heat storage systems require large storage tanks due to lower energy density compared to other technologies. This technology is fully commercialized on the industrial scale mainly due to low capital cost compared to other heat storage methods. But, it requires isolation systems to prevent thermal losses over time.

11. How does latent heat storage work? Latent heat storage is based on the amount of energy (heat) needed to change the material phase, for example, from solid to liquid or liquid to gas. The temperature range is between 20 and $80\,°C$, and the widely used phase-change materials are organic compounds and inorganic salts, such as water, molten salts, or sodium hydroxide. During the phase change, the heat forms or breaks the bond between the molecules that can absorb or release a large amount of heat.

12. What are the main components of latent heat storage? The storage medium or phase-change material (PCM) and heat exchanger for transferring heat to the medium (Figure 5.2).

13. What are the benefits and drawbacks of latent heat storage technology? Some of the benefits of latent heat storage technology include: (i) storing and releasing heat at constant temperatures, (ii) high energy density storage (50–$100\,kWh/m^3$), which requires smaller storage volumes, (iii) high efficiency (up to 90%), and (iv) medium capacity (50–$100\,kWh/t$). This technology has various applications, especially for storing heat for several hours

or days. The main drawbacks of this technology include: (i) low heat transfer rate (or thermal conductivity), (ii) high cost, (iii) limited storage periods, (iv) thermal losses, and (v) large temperature difference requirements between the storage material and heat source. The storage medium is one of the major cost drivers of this technology. Also, using fossil fuel-based mediums (e.g., paraffin wax) has negative environmental impacts from manufacturing and disposing of them. Renewable mediums (e.g., salt hydrates) have fewer environmental impacts.

14. What is aqueous salt solution? and how does it work? Aqueous salt solution is a type of latent heat storage that works by heating the storage medium (a mixture of water and salt) to a high temperature. The solution can reduce the freezing point and remain liquid at high temperatures. Then it can be used to store and release heat when needed to generate steam and power for various applications. The main components of this technology are the storage medium, tank, and heat exchanger. The salt concentration in the solution is one of the leading performance factors.

15. What are the benefits and drawbacks of the aqueous salt solution for heat storage? The benefits are high efficiency, long lifetime, low cost, and high applications. The main drawback is the high-temperature requirements for the highly efficient operation.

16. How does thermochemical heat storage work? It is based on reversible reactions (endothermic and exothermic) that can be either physical or chemical phenomena (Figure 5.3). During the endothermic reaction (charging), an element decomposes into two elements, using heat and storage separately. During the exothermic reaction (discharging), the heat is released from bonding the elements.

17. What are the main types and key attributes of thermochemical heat storage? There are two main thermochemical heat storage: chemical reactions and sorption-based processes. This technology has high energy density storage ($100-700\,kWh/m^3$), high efficiency (up to 99%), and high capacity ($120-250\,kWh/t$).

18. What are the benefits of thermochemical heat storage technologies? High energy storage density and low energy losses over a long storage period, which make them well suited for long-term energy storage compared to latent and sensible heat storage systems.

19. How does sorption-based heat storage work? Sorption-based heat storage uses thermochemical principles for storing energy, which are endothermic and exothermic or reversible chemical reactions to store and release thermal energy (Figure 5.4). The process can be repeated several times and release and store heat during adsorption and desorption, respectively.

20. What are the main components and types of sorption-based heat storage? The key components are sorbent materials, the reactor and heat exchanger, a vapor transport system, and an energy storage and retrieval system. The two main types of thermochemical sorption systems are adsorption and absorption, using sorbent materials with specific chemical compositions (e.g., water). Adsorption is a surface interaction between a gas and a solid. Particularly, gas is adsorbed on the sorbent materials that cause chemical reactions and release heat or thermal energy. The desorption process includes heating or lowering the pressure to release gas or liquid that causes sorbent material to cool down and store heat. Absorption is the volume interaction between the gas or liquid (e.g., water) and another liquid (sorbent materials), such as salts and clays.

21. What are the benefits and drawbacks of sorption-based heat storage technology? This technology gained attention due to low heat loss and high energy storage density, as well as its broad applications for space heating, hot water supply, short- and long-term energy storage, and industrial heat and cooling systems. Additionally, sorption-based heat storage systems do not release any emissions during the operation and have low environmental impacts from the production and disposal of the required resources. Some drawbacks are high cost, complex design, and sorbent material degradation over time. Also, sorbent materials can be toxic and hazardous and need special processes for production and disposal.

22. How does chemical-based heat storage work? Chemical-based heat storage can store thermal energy using chemical substances and reversible chemical reactions (store and release). The science behind chemical reactions involves forming and breaking chemical bonds to release and store heat in the form of chemical potential energy.

23. What are the main components and factors of chemical-based heat storage technology? The key components of chemical-based heat storage are the reactant, reactor, heat exchange, and energy storage system. The main factors on the performance are energy storage capacity, depending on the specific chemical reaction, and the energy balance depending on the heat input and output. Other key factors are reactant type, reactor temperature, and time.

24. What are the main types of chemical-based heat storage technology? Chemical-based heat storage can be classified into three main types based on the reaction mechanism, which are gas–gas, liquid–gas, or solid–gas reactions. For example, gas–gas reactions can use ammonia (NH_3) to store heat by converting it to nitrogen and hydrogen. The reaction can be faster, using different catalysts and under high pressure.

25. What are the benefits and drawbacks of chemical-based heat storage technology? The benefits of chemical-based heat storage technologies are high

energy storage density and long-term storage capacity, but they have a high cost due to reactants and complex reaction systems. Also, the reactants can be non-renewable and toxic, posing environmental impacts and safety concerns.

Chapter 6

1. What are the main types of biofuels? Biofuels can be classified into two main types: gaseous (e.g., biogas and biomethane) and liquid hydrocarbon fuels (e.g., bioethanol and biodiesel). Gaseous biofuels have been used as clean cooking fuels for power and heat generation. Liquid biofuels have been used in road transportation that can be expanded to aviation and shipping. Liquid biofuels are mainly produced from crops (e.g., corn, sugarcane, and soybeans) that can compete and interrupt food production, and limit the supply for biofuel production.

2. What are the most common biofuel production processes? Check Figure 6.1.

3. What is biogas? Biogas is a mixture of different gases (e.g., CO_2, CH_4, H_2O, and H_2S) produced from the anaerobic digestion of organic wastes (e.g., food wastes, manure, or sludge) and biomass feedstocks (e.g., forest or agriculture residues) in the absence of oxygen. CH_4 in biogas is relatively high (50–75%) that can be used for heat and power generation. Biogas energy content is similar to that of natural gas since CH_4 is the primary component of natural gas.

4. What is anaerobic digestion? Anaerobic digestion is a chemical process in which microorganisms and bacteria digest (break down) organic wastes in oxygen-free, sealed vessels or reactors (Figure 6.2). This technology can produce biogas and nutrient-rich fertilizers and cut up to 10 gigatons of GHGs.

5. What are the benefits and drawbacks of biogas production? The benefits include: (i) generating revenue from waste products, (ii) reducing GHG emissions, and (iii) providing renewable energy sources and byproducts (e.g., fertilizers) from its digestate (leftover materials from the anaerobic digestion process). The major drawbacks are: (i) high capital and operational costs and (ii) uncertainties about biogas quality due to inconsistency in biomass feedstocks.

6. What are the main steps in the anaerobic digestion process? And what are the main factors? Hydrolysis (large molecules to small ones), acidogenesis (small molecules to organic acids), acetogenesis (organic acids to acetate), and methanogenesis (acetate to CH_4 and CO_2). The main parameters that can impact the process performance are organisms type, loading rate, retention time, pH, and temperature.

7. What are the main parameters and challenges in algae anaerobic digestion? The key parameters impacting the process yield are temperature, pH, algae

loading rate, and process time. In this process, the temperature range is 25–40 °C, pH range is 6.5–8.5, and the process time is around 20–30 days. The main challenge of algae anaerobic digestion is the inconsistency in algae components and compositions (e.g., lignin type, cellulosic fibers, and polyphenols) that can limit digestibility and reduce process yield. Additionally, seasonal growth and location are other associated problems.

8. What is biomethane? Methane-based energy from biomass (biomethane) is a low-emission solution that captures emissions and generates energy from organic wastes in landfills instead of using fossil fuels (e.g., coal and natural gas). Biomethane production is mostly achieved through the upgrading of biogas (removing CO_2 and other gases) to pure methane, or gasification and anaerobic digestion of various feedstocks, such as municipal solid wastes, manure, or agriculture residues.

9. What are the most used biomethane production technologies? Check Figure 6.4.

10. What are the main steps in the gasification process? Gasification process involves four steps: (i) drying (reducing the moisture content in the materials), (ii) pyrolysis (breaking down materials into small molecules), (iii) combustion (producing heat and gases in partial oxygen present), and (iv) reduction (removing CO_2 and other impurities, using reduction agent).

11. What are the main types of gasifiers? Gasifiers have been used for centuries and can be classified into three main types based on the materials' contacting method with gasification reagents (heat and air), which are moving-bed (e.g., updraft or downdraft), fluidized-bed, and entrained-bed reactors.

12. What are the key parameters for evaluating the gasification process? Operating conditions (e.g., temperature and pressure), process capacity (e.g., feed rate and particle size), syngas composition before and after cleaning, power output, total products throughput, emissions, water consumption in the cooling tower, and contaminants in wastewater.

13. What are the applications of syngas from the gasification process? Gasification syngas (a mixture of CO and H_2) can be used for the following: (i) power generation, using gas turbines and integrated coal gasification combined cycle or steam power and Rankine cycle; (ii) hydrogen generation, using fuel cells; (iii) production of methanol or similar chemicals, such as acetic acid, polyolefins, methyl esters, and formaldehyde, (iv) productions of Fischer–Tropsch process products, such as naphtha, wax, diesel, and gasoline, and (v) production of other products, such as ethanol and synthetic natural gas (SNG).

14. What are the chemical reactions during the gasification process of organic materials (e.g., woody biomass or agriculture waste)? (i) $C_6H_{12}O_6 + 6O_2 \rightarrow 6CO_2 + 6H_2O$ and (ii) $C_6H_{12}O_6 \rightarrow 3CH_4 + 3CO_2$.

15. What are the steps for biomethane production through gasification with catalytic methanation pathway? Syngas from the gasification process contains CO, H_2, CH_4, and CO_2. The catalytic methanation process can convert CO and H_2 to methane and water ($CO + 3H_2 \rightarrow CH_4 + H_2O$), which can increase the methane content of syngas. The chemical reaction formula for gasification and catalytic methanation from biomass is as follows: $C_6H_{12}O_6 + 6O_2 + 3CO + 9H_2 \rightarrow 3CH_4 + 6CO_2 + 9H_2O$.

16. What are the steps for biomethane production through gasification with biological methanation pathway? Gasification with biological methanation pathway combines the benefits of both thermochemical and chemical conversion processes for renewable biomethane production from waste streams. Biological methanation of syngas from gasification is an anaerobic process in which bacteria break down syngas into methane and CO_2 using various microbial groups, such as methanogenic archaea. The biological methanation has several benefits compared to the catalytic methanation process, such as milder operation conditions (e.g., low temperature and pressure) and less sensitivity to syngas purity and carbon/hydrogen ratio.

17. What are the mature bioethanol production technologies? Check Figure 6.6.

18. What are the main steps during gasification syngas fermentation of lignocellulosic for bioethanol production? It involves three main steps: (i) biomass gasification, (ii) syngas fermentation, and (iii) bioethanol distillation (Figure 6.7). During biomethane production, air, oxygen, or steam can be used as a gasifying agent. Then syngas from gasification is fed to the fermenter to convert CO and H_2 to bioethanol and other byproducts via microorganisms (e.g., yeast or bacteria). After fermentation, bioethanol can be purified through a distillation process, including heating and condensing steps.

19. What are the chemical reactions during gasification syngas fermentation of lignocellulosic? (i) Gasification: lignocellulosic biomass ($C_6H_{10}O_5$) $+ 2H_2O + 2O_2 \rightarrow 6CO + 10H_2$ and (ii) syngas fermentation: $4H_2 + 2CO \rightarrow CH_3CH_2OH + 2H_2O$.

20. What are the main steps during enzymatic fermentation of lignocellulosic for bioethanol production? Enzymatic fermentation is a common method for converting lignocellulosic feedstocks (e.g., woody biomass) into bioethanol through microorganisms (e.g., yeast and bacteria) that can metabolize complex sugars, break them down to simpler sugars, and produce bioethanol. The chemical process includes pretreatment and hydrolysis to release and break down sugars, which are fermented by microorganisms to bioethanol and lignin (Figure 6.8). Particularly, enzymes help breaking down the hemicellulose and cellulose in lignocellulosic feedstocks to sugars during hydrolysis.

21. What are the advantages and disadvantages of enzymatic fermentation of lignocellulosic for bioethanol production? This production pathway can reach up to 80% process yields. The drawbacks are pretreatment requirements, high water use for hydrolysis, and high processing costs due to enzyme requirements. Also, bioethanol production from lignocellulosic feedstocks (e.g., wood and grass) requires more processes than from starch-based crops.

22. How does enzymatic fermentation convert sugar and starch to bioethanol? The process uses enzymes (e.g., alpha-amylase and glucoamylase) to convert complex carbohydrates or disaccharides from starch or sucrose, respectively, into simple sugars (e.g., maltose, fructose, and glucose) and then fermented by yeast to produce bioethanol. The chemical reaction of fermentation of glucose is as follows: $C_6H_{12}O_6 \rightarrow 2C_2H_5OH + 2CO_2$.

23. What are the benefits and drawbacks of enzymatic fermentation for bioethanol production? The benefits include: (i) higher efficiency than conventional fermentation methods due to the use of enzymes, (ii) lower energy consumption and temperatures, (iii) higher process yields compared to conventional fermentation methods, (iv) lower environmental impacts and emissions, and (v) lower processing costs due to higher yield and efficiency. The drawbacks are: (i) high enzymes cost, (ii) process complexity due to enzyme sensitivity to temperature and pH changes, and (iii) longer production time that can reduce the capacity.

24. What are the physical and chemical characteristic differences between biodiesel and diesel #2? Check Table 6.2.

25. What are the benefits and drawbacks of biodiesel production? The benefits of biodiesel compared to other biofuels include ease of use, low GHG emissions, stability in extended storage, and improved engine operation due to increased cetane number. The drawbacks are higher production costs compared to diesel, limited feedstocks, compatibility issues with existing engines and infrastructure, and environmental impacts due to high land use and resources.

26. What are the mature technologies for biodiesel production? There are several methods for biodiesel production from different feedstocks, and the main methods use gasification, pyrolysis, transesterification, or liquefaction technologies (Figure 6.9).

27. How does esterification convert fatty acids to biodiesel? Esterification process converts free fatty (carboxylic) acids in vegetable oils or animal fats into methyl esters in the presence of catalysts (Figure 6.10). This process uses short-chain alcohols (e.g., methanol) and catalysts (e.g., potassium hydroxide or sodium hydroxide) to convert oil to biodiesel and glycerin, which is a co-product of the esterification process.

28. What are the key parameters for biodiesel production through esterification of fatty acids? The key parameters are water content, reaction temperature and time, and alcohol-to-acid mixing ratio. Reducing water content can increase the esterification reaction and conversion rates. The reaction temperature is around 55–60 °C, and it takes several hours to complete the esterification process. The longer process can increase the process yield; however, very long processes can lead to side reactions that form other byproducts.

29. What are the steps during the hydrogenation of vegetable oil for biodiesel production? The steps include various processes, such as triglyceride hydrogenation to decompose vegetable oil to monoglycerides, diglycerides, and carboxylic acids, and then convert them into alkanes via decarboxylation and hydrogenation at high pressure (over 400 psi) and temperature around 350 °C using various catalysts for hydroprocessing.

30. How does the microbial lipid fermentation process produce biodiesel? The microbial lipid fermentation process uses microorganisms (e.g., algae, yeast, or bacteria) to convert sugars (e.g., glucose or sucrose) or other carbon sources into lipids and hydrocarbons that can be upgraded through a complex chemical (metabolic) reaction (e.g., hydroprocessing) to synthetic iso-paraffin or biodiesel. Generally, lipids are made of glycerol and fatty acid chains, consisting of a hydrocarbon chain with a carboxylic acid group. The critical parameters are conversion process configuration (e.g., temperature and pH), microorganism type, and sugar type.

31. What are the benefits and drawbacks of the microbial lipid fermentation process for biodiesel production? The benefits include: (i) high conversion efficiency for using microorganisms to produce synthetic iso-paraffin from a wide range of carbon sources (e.g., plant, food, and agriculture residues), (ii) lower GHG emissions due to the use of renewable carbon sources compared to traditional products with similar characteristics, and (iii) variety of applications, such as fuel additive to improve diesel performance, solvent for industrial adhesives, and lubricant for high-performance engines. The drawbacks are high production costs and scale-up challenges.

32. What are the steps and key parameters during the alcohol-to-jet fuel process? Generally, the conversion process includes fermentation and upgrading, and the key parameters are biomass and microorganism types, and reactor design. Alcohol-to-jet fuel (aviation biofuel) process involves the conversion of biomass feedstocks (e.g., lignocellulosic materials or organic matter derived from plants and animals) through biochemical processes into a longer chain of hydrocarbons and alcohols. The biochemical processes use microorganisms (e.g., bacteria or yeast) to break down the biomass into fatty acids and alcohols. Then the alcohols can be further processed through dehydration,

oligomerization, and hydrogenation processes to produce jet fuel or biodiesel (Figure 6.11).

33. What are the benefits and drawbacks of the alcohol-to-jet fuel process? The benefits are similar to microbial lipid formation, such as lower emission, higher yield, and improved waste management. The main drawbacks are complex chemical reactions, process efficiency and optimization challenges, and total costs compared to other biofuels.

34. What are the steps for biofuel production through pyrolysis and upgrading processes? Pyrolysis is a thermochemical process (similar to gasification) that can break down organic materials rapidly (in a few seconds), such as biomass feedstocks to pyrolysis oil, pyrolysis char, and pyrolysis gas under high temperatures (400–700 °C) in the absence of oxygen. Pyrolysis oil can be upgraded using different technologies (e.g., deoxygenation, hydrogenation, or catalytic cracking) to biofuels, e.g., biodiesel (Figure 6.12). The chemical reactions during the conversion of biomass feedstocks are complex and depend on biomass type, process condition (e.g., temperature and pressure), reactor design (e.g., free fall or fluidized bed), and catalyst types.

35. What are the steps for biofuel production through hydrothermal liquefaction and upgrading processes? Hydrothermal liquefaction is a thermochemical process (similar to pyrolysis) that can deconstruct high moisture content (wet) feedstocks (e.g., algae) to bio-oil under high pressure and moderate temperature (200–400 °C). Similar to other thermochemical processes, the chemical reactions depend on the biomass type, process configurations, and catalyst type. The main product of hydrothermal liquefaction is bio-oil that can be upgraded using different processes, such as hydrodeoxygenation or catalytic cracking with catalysts for removing unwanted compounds and producing compatible biofuels (Figure 6.13).

36. What are the chemical reactions for biofuel production through gasification-FT with and without CCUS? Gasification-FT without CCUS: $nCO + nH_2 \rightarrow C_nH_n + nH_2O$. Gasification-FT with CCUS: $nCO + nH_2 + nCO_2 \rightarrow C_nH_nCOOH + nH_2O$.

37. What are the steps for biofuel production through gasification-FT mixed with hydrogen enhancement? Gasification-FT mixed with hydrogen enhancement is a thermochemical pathway for biodiesel production, using several processes, such as gasification, syngas conditioning, FT, and hydrotreatment (Figure 6.14). Gasification can convert organic materials to syngas at high temperatures. Syngas is a mixture of several gases, such as CO and H_2. Then FT process can convert syngas to liquid hydrocarbon fuels, and the hydrogen enhancement process can improve the quality of the hydrocarbon fuels, using additional hydrogen. Particularly, hydrogen enhancement uses different

methods, such as hydrodeoxygenation, to remove oxygen and increase hydrogen to produce a fuel similar to fossil-based fuels (e.g., diesel).

38. What is the transesterification process? Transesterification is a low-temperature chemical process that uses alcohol in the presence of a catalyst to open up the physical structure of feedstocks (e.g., microalgae) and make sugar polymers (e.g., hemicellulose and cellulose) available for producing biodiesel or fatty acid methyl esters and byproducts, such as glycerin. The general chemical reaction is as follows: microalgal lipids (triglyceride) + alcohol → biodiesel + glycerol.

39. What are the steps for biodiesel production from microalgae through the transesterification process? The microalgae transesterification conversion process includes lipids extraction and purification, catalysis, and biodiesel separation and purification from glycerol (Figure 6.15).

40. What is the hydrotreating conversion process? Hydrotreating is a hydrogenation process that can convert high-moisture-content feedstocks (e.g., microalgae) under high temperature and pressure in the presence of catalysts to hydrocarbons (Figure 6.16), and the overall chemical reaction is as follows: Triglyceride + nH_2 → hydrocarbon + nH_2O.

41. What are the main CCUS strategies? Absorption, adsorption, membranes, and cryogenic processes.

42. What are the primary carbon-capturing methods? Pre-combustion (carbon is removed before combustion), post-combustion (carbon is removed after combustion), and oxy-fuel combustion (fuel is combusted with pure oxygen). Between these methods, pre-combustion can be a cost-competitive CCUS strategy for power and chemical production from high-pressure, oxygen-blown gasification processes (Figure 6.17).

43. What are the challenges of using CCUS strategies? (i) reducing energy requirements for carbon capturing and (ii) minimizing preparation requirements (e.g., pressure, target purity, temperature, and flow rate) to prevent carbon release to the environment by finding a suitable end use (e.g., enhanced oil recovery).

Chapter 7

1. What are mature low-emission hydrogen production technologies? Check Figures 7.1 and 7.2.

2. What are the main demands of hydrogen? Hydrogen is used mainly for ammonia production in the fertilizer industry, methanol production in the chemical industry, and oil refinery in the transportation industry.

3. What are the main methods for large-scale hydrogen production? And what are the key challenges? Natural gas autothermal reforming and methane

steam reforming, and the key challenges are the production cost, handling, and safety aspects.

4. What are the critical parameters for hydrogen production via thermochemical processes? Feedstock type (e.g., water or methane), conversion reactor (e.g., fluidized- or fixed-bed), catalyst, required energy and heat source (e.g., nuclear, renewable, or fossil fuels), as well as separation and purification methods.

5. What is the most common method for hydrogen purification to achieve the required purity for the commercial scale? Pressure swing adsorption is the most common commercial method to achieve the required hydrogen quality, using a molecular sieve by removing other gases (e.g., nitrogen, oxygen, CO, and H_2S).

6. What are the main steps of the pressure swing adsorption process? Adsorption, co-current depressurization, purge, and counter-current depressurization.

7. What are the pros and cons of hydrogen production through thermochemical processes mixed with CCUS? The main benefits of thermochemical processes mixed with CCUS include high efficiency, low emissions, and feedstock flexibility. The main drawbacks are high requirements (e.g., energy, temperature, and pressure) and high capital and operational costs. The environmental impacts are highly dependent on feedstock type and energy source.

8. What is methane steam reforming? Methane steam reforming is a mature process that can generate hydrogen from methane, using high-temperature steam (700–1,000 °C). In this process, methane reacts with steam under high pressure (up to 360 psi) in the presence of catalysts.

9. What are the main components of methane steam reforming? (i) the reformer for steam and methane reaction, (ii) heat source for the endothermic process, (iii) the catalyst and its support (e.g., nickel on alumina), and (iv) the separation processes, such as pressure swing adsorption.

10. What are the key parameters of the methane steam reforming process? Reaction temperature and pressure, catalyst performance, and steam and carbon ratio.

11. What are the benefits and drawbacks of the methane steam reforming process? The benefits of methane steam reforming are reliability and high efficiency for large-scale hydrogen production from natural gas, biogas, and coal. The byproducts of this process (e.g., CO) can be used in different applications to increase efficiency and reduce costs. The main drawbacks include: (i) GHG emissions (e.g., CO_2, SO_2, and NOx) and environmental impacts of using fossil fuel-based resources (e.g., natural gas or coal), (ii) energy-intensive processes, and (iii) expensive separation processes. The major cost drivers are feedstock cost, energy use, and separation costs.

12. What are the differences between carbon capturing before and after hydrogen separation in methane steam reforming? The main benefits of carbon capturing before hydrogen separation (like pre-combustion carbon removal) are high CO_2 concentration and low capital costs, but the main drawback is the low CO_2 capturing rate (up to 60%). The key benefit of carbon capturing after hydrogen separation (like pre-combustion carbon removal) is high capturing rare (around 90%); however, it requires high capital costs. The latest techno-economic studies show that the levelized cost of hydrogen production increases between 18 and 45% when the steam reforming process is integrated with CCUS technologies.

13. What is natural gas autothermal reforming? Natural gas autothermal reforming uses a combination of reforming processes (e.g., steam reforming, partial oxidation, and autothermal reforming), along with other processes such as purification and cleanup for removing impurities. The autothermal reforming process converts natural gas to syngas (e.g., H_2, CO, and CO_2) through oxidizing natural gas, using oxygen in the presence of a catalyst (e.g., nickel-based materials).

14. What are the key parameters for improving the output and efficiency of natural gas autothermal reforming? (i) reactor design and configuration (e.g., temperature and pressure), (ii) natural gas and oxygen flow and mixing rates, and (iii) catalyst type, bed design, and composition.

15. What are the benefits and drawbacks of the autothermal reforming process? The main benefits are: (i) cost-effective method compared to other methods (e.g., gasification or electrolysis), (ii) high energy efficiency (up to 80%), and (iii) high purity H_2 level that can address several industrial needs. The main drawbacks are: (i) the high capital cost of the oxygen production process, which can be addressed in large-scale hydrogen production; (ii) an energy-intensive process; and (iii) high GHG emissions and environmental impacts that can be addressed by mixing this process with CCUS strategies. The major cost drivers are natural gas price and energy use.

16. What are the main differences between natural gas autothermal reforming and methane steam reforming? (i) autothermal reforming process uses both oxygen and steam, but methane reforming uses only steam; (ii) autothermal process has higher temperatures (up to 1,000 °C) and pressures (up to 30 bar); (iii) autothermal process produces H_2, CO, CO_2, and water, but methane reforming produces H_2, CO_2, and a small amount of CO; (iv) autothermal process is more energy efficient, has a higher yield, and requires less heat, (v) autothermal process requires catalysts for both oxidation and steam reforming, but methane reforming needs catalysts only for steam reforming reaction, and (vi) it is easier to mix CCUS strategies with autothermal process due to the gas mixture, such as higher CO_2 concentrations. In summary, the

autothermal reforming process is more complex and requires more energy and catalyst; however, it has higher hydrogen yields and effective CCUS integration.

17. What are the main feedstock types for hydrogen production? The two main types of biomass feedstocks for hydrogen production are energy crops (e.g., poplar and switchgrass) and biogas from organic residues (e.g., animal wastes and municipal solid wastes) after the anaerobic digestion process.

18. What are the benefits and drawbacks of hydrogen production through biomass gasification? The main benefits include: (i) biomass and waste management, (ii) lower environmental impacts compared to fossil fuels due to the use of renewable and local resources, (iii) a flexible and diverse source of hydrogen from a variety of feedstocks, and (iv) the need for high CO_2 storage. The major drawbacks are feedstock availability throughout the year, pretreatment processes (e.g., size reduction), syngas cleaning to reduce impurities (e.g., tar, acids, and particulates), and the high cost of storing and transporting large quantities. The major cost drivers are feedstock (over 80%), capital cost (around 12%), and operation cost (around 5%), including energy use, gas cleaning and condition, and hydrogen purification and compression.

19. What are the benefits and drawbacks of hydrogen production through methane pyrolysis? The benefits are: (i) low GHG emissions, (ii) non-catalytic process, (iii) no requirements for purification steps due to high purity hydrogen, and (iv) the potential to integrate with CCUS strategies for carbon production. The drawbacks include tar formation, high energy use, and high capital and operational costs.

20. What are the key components of electrolysis processes? Anode, cathode, membrane, and electrolyte.

21. What are the benefits and drawbacks of hydrogen production through water electrolysis? The key benefits are simplicity, low temperature, zero-carbon emissions, high hydrogen purity, and oxygen byproducts. The drawbacks are process requirements (e.g., high pressure), energy storage challenges, low efficiencies, and high capital costs.

22. What are the key parameters affecting the process efficiency of alkaline electrolysis? Electrolyte concentration, current density (current amount per unit area), temperature, and pressure. For example, a higher current density can increase the hydrogen production rate, but it can increase energy use and reduce the electrolysis cell lifespan. Also, the process pressure depends on the used electrolyte and cell design.

23. What are the benefits and drawbacks of alkaline electrolysis for hydrogen production? The benefits include low capital costs due to low material costs (compared to other electrolysis technologies), long lifetime, high hydrogen production, compact design, and high energy efficiency (over 80%). The

drawbacks include low hydrogen purity, low partial load range, low current density, and complex KOH (potassium hydroxide) handling.

24. What are the benefits and drawbacks of the proton-exchange membrane electrolysis process? The benefits are high proton conductivity, high efficiency, high hydrogen purity, modular design, low process footprint, pure water solution (no alkaline), and fast response. The drawbacks are high material costs, supply chain challenges for precious metals, shorter lifespan than alkaline electrolysis, limited capacity, and low durability. The major cost drivers are bipolar plates (51%) and membrane electrode assembly manufacturing (10%).

25. What are the key parameters affecting solid oxide electrolyzer performance? Process temperature and pressure, current density, and electrolyte composition. Particularly, high temperature and pressure can increase the reaction rates and efficiency. In this process, current density and electrolyte composition can affect the electrochemical reactions, ion conductivity, and stability.

26. What are the benefits and drawbacks of solid oxide electrolyzers? The main benefit is high efficiency (up to 80%), flexibility to produce hydrogen and power as fuel cells, and a long lifespan. However, the drawbacks are high cost, high-temperature requirements, and carbon deposition that can decrease process efficiency. The solid oxide electrolyzer has lower maturity compared to alkaline and polymer exchange membrane electrolysis processes due to the operation requirements (e.g., high temperature). The extra heat can be reused in other processes.

27. What are the challenges and strategies for improving solar-based thermochemical hydrogen production? The main challenges are the efficiency and durability of the thermochemical reactor design (e.g., heat cycling) or reactant materials (e.g., cerium oxide or copper chloride). Other challenges are significant capital costs and a large area for efficient concentration, which is why this technology has low TRL and requires more investigation. Some of the strategies to address the existing barriers are: (i) improving reactant materials to optimize heat transfer and durability, (ii) reducing heat losses, (iii) improving the membrane materials to optimize conductivity and durability, (iv) improving efficiency by optimizing the electrolysis process, and (v) improving the thermochemical process by developing thermal and chemical storage systems.

Chapter 8

1. What are mature hydrogen-based storage technologies? Check Figure 8.1.
2. What are the main technical parameters for hydrogen storage? (i) Weight, volume, and discharge rate; (ii) heat requirements; (iii) capital and operational costs; and (iv) storage capacity and recharging time.

3. What are the main forms of hydrogen storage? Gas, liquid, and solid forms.
4. What does cryogenic hydrogen storage mean? And what are the benefits? Hydrogen gas cooled to near cryogenic temperatures (cryogas) is another way of storing hydrogen due to its high volumetric energy density. Cryogenic storage involves hydrogen storage as a supercritical fluid at low temperatures that can provide high storage capacity with high energy density, but it needs a cooling system.
5. What are the main challenges for solid hydrogen storage? The lack of sufficient reversibility and the complex hydrogen extraction process.
6. What are the key challenges of physical-based hydrogen storage? The proper discharge temperature and the ability to recharge the system quickly (a couple of minutes).
7. What are the major parameters for using hydrogen storage tanks? (i) hydrogen properties (density, volume, and amount), (ii) material and design compatibility (resistance and external damage), (iii) temperature and pressure management for leakage and relief systems, (iv) Safety measures (flammability and risks), and (v) storage methods (gas, liquid, or solid). Understanding these parameters helps develop efficient tanks with various application capabilities.
8. What are the main physical-based hydrogen storage forms? Check Figure 8.2.
9. What are the main advantages of lightweight composite hydrogen storage tanks? Lightweight, corrosion resistance, size flexibility, gunfire and impact safety, abrasion resistance, and cost-competitiveness.
10. What are the main drawbacks of composite tanks for hydrogen storage? (i) large size and physical volume, which makes it difficult to adjust it with the available spaces, (ii) high cost (over \$500 per kg), (iii) high-pressure requirements for compressing the gas, and (iv) safety challenges due to high pressure. These composite materials require further studies under various conditions (e.g., temperature and pressure) to assess the long-term effects.
11. What are the benefits and drawbacks of hydrogen storage with glass microspheres? The benefits are: (i) a safer method for hydrogen storage in low-pressure compared to other methods, (ii) low container costs, and (iii) hydrogen storage density of 5.4 wt%. The key challenges with glass microspheres are: (i) low volumetric density, (ii) high pressure and temperature requirements for filling, and (iii) hydrogen leaks at ambient temperature.
12. What are the pros and cons of liquid hydrogen storage compared to gas storage? Liquid hydrogen has a better storage density at relatively low pressures compared to compressed gas; however, it takes 30–40% extra energy to produce liquid hydrogen and requires super-insulated containers.
13. What are the most common liquid hydrogen storage methods? Cryogenic, borohydride ($NaBH_4$) solutions, and rechargeable organic liquids.

14. What are the mature underground storage methods? Salt caverns, depleted oil and gas reservoirs, and deep aquifers. Check Figure 8.4.

15. What are the benefits and drawbacks of underground hydrogen storage? (i) high capacity (TWh for weeks or months) compared to aboveground storage sites with limited capacity (MWh for hours or days), (ii) lower storage costs, and (iii) high safety and less vulnerability to terrorist attacks. However, underground hydrogen storage involves gas leakage and losses due to faults, fractures, or biological reactions, which requires a leakage test with a maximum allowable leakage rate of $160 \, m^3$/year.

16. What is the salt cavern method for hydrogen storage? Salt caverns are mainly cylindrical pits in thick underground salt deposits, and depend on geological conditions (e.g., salt mechanical properties, tightness, and resistance to chemical reactions). Salt cavern storage has been used for large-scale hydrogen storage for medium to long periods, enabling seasonal hydrogen gas storage with up to 10 times injection and extraction per year. This method has gained high interest due to its stability, ease of management, and high volume capacity of up to 1 million m^3 at a maximum of 200 bar pressure. However, this technique has several technical challenges, such as the installation, transfer capacity, and tightness of boreholes, as well as location and environmental limitations.

17. What is a depleted oil and gas reservoir for hydrogen storage? They are geological traps with impermeable layers, supported by aquifers. These reservoirs are known options for hydrogen storage due to their reliable tightness and geological structures.

18. What is the aquifer method for hydrogen storage? And what are the pros and cons of this method? Aquifers are permeable, porous media that can store gas or water in their pore space. The main parameters for using aquifers for hydrogen storage are rock structure and properties, and impermeable layers for avoiding gas migration. This method enables seasonal hydrogen gas storage with a few annual injection and withdrawal cycles. Due to greater geological distribution, this method is much more flexible for large-scale hydrogen storage than salt caverns. The capacity of aquifers can be determined by their geological structure (e.g., reservoir rocks, size, and thickness). Some barriers are: (i) uneven distributions and complexity, (ii) lack of reliable storage capacity, and (iii) high costs of finding reliable storage sites.

19. What is material-based hydrogen storage? Material-based storage is another approach for storing hydrogen, mainly in the solid form, using different materials to improve critical factors, such as volumetric and gravimetric capacities, required storage pressure and temperature, and cost.

20. What are the mature material-based hydrogen storage methods? Metal hydrides, chemical hydrogen, and adsorbents. Check Figure 8.6.

21. What are the suitable materials for solid hydrogen storage? (i) carbon-based materials (e.g., activated charcoal, nanotubes, and graphite nanofibers) or high surface area materials (e.g., zeolites, clathrate hydrates, and metal oxides), (ii) water-reactive chemical hydrides (e.g., NaH, CaH_2, MgH_2) and thermal chemical hydrides (e.g., AlH_3 and NH_3BH_3), and (iii) rechargeable hydrides (e.g., alloys and nanocrystalline).

22. What are the benefits of solid storage compared to gas or liquid storage? Lower volume and pressure, and higher hydrogen purity.

Glossary

6R	Reduce, reuse, recycle, recover, redesign, and remanufacture
Absorption	Capturing or storing within solids
Adsorption	Capturing or storing on the surface of solids
Anthropogenic	Human-caused emissions
Biogenic	Natural-made emissions
Carbon budget	The total emissions (including from GHG and land-use change) that can be added without exceeding the target temperatures. Based on the IPCC, the budget limit for the 2 °C target is 1,000 billion tons of CO_2 by 2,100
Carbon offsetting	Emission reduction strategies for companies that cannot change their entire operations and reduce their emissions to zero due to the lack of technologies or pathways to remove emissions at a meaningful scale, such as fossil fuel producers, airlines, and steel manufacturers
Capacity factor	The percentage of time the plant operates at full capacity
Combined heat and power (CHP)	It means generating heat and electricity from the same energy source or process with single or different energy inputs
Cryogas	An extremely cold compressed gas in cryogenic temperatures (between −150 and −273 °C)
Cryogenic process	The cryogenic process refers to cooling down hydrogen to liquid form with critical pressure of around 13 bar and a temperature below −253 °C, as well as a volumetric and gravimetric storage density of approximately 70 kg/m^3 and 20 wt%, respectively
Decarbonization	The process of reducing carbon emissions
Diversion	Channeling water of a river through a canal or pipe

Net-Zero and Low Carbon Solutions for the Energy Sector: A Guide to Decarbonization Technologies, First Edition. Amin Mirkouei.
© 2024 John Wiley & Sons, Inc. Published 2024 by John Wiley & Sons, Inc.

Endothermic	A process that requires heat
Energy balance	It refers to the balance between energy input (usually from the sun) and energy output (such as heat radiation) within an ecosystem
Enthalpy	A thermodynamic property of a system that describes the total heat content of a system at constant pressure
Entropy	The measure of a system's thermal energy per unit temperature that is unavailable for doing useful work
Exothermic	Release heat
Greenhouse gas (GHG) emissions	CO_2, CH_4, N_2O, CFC, and HFC
Hydrolysis	A chemical process that involves the breaking down of a compound using water molecules
Impoundment	Storing water in a reservoir
Near-zero solutions	Solutions with lower GHG emissions compared to traditional solutions
Net-negative target	Remove more emissions than they emit
Net-zero target	Carbon neutral or do not emit GHG emissions during generation, storage, and consumption
Petawatts	1 million gigawatts
Polysilicon	Polycrystalline silicon
Power capacity	Power capacity (kW) refers to the amount of power the storage technology can produce during the discharging process. The power density (kWh/L) refers to the ratio of power capacity to storage system capacity
Syngas	Synthesis gas
Therm	Approximately 100,000 BTUs
Technology readiness levels (TRL)	A measure for evaluating a technology's maturity from basic research to commercialization, and a higher number indicates that the technology is closer to commercialization in the market.

Index

Net-Zero and Low Carbon Solutions for the Energy Sector: A Guide to Decarbonization Technologies, First Edition. Amin Mirkouei.
© 2024 John Wiley & Sons, Inc. Published 2024 by John Wiley & Sons, Inc.